*Construction
Foreman's Job
Guide*

Construction Foreman's Job Guide

James E. Clyde, P. E.

A Wiley-Interscience Publication

JOHN WILEY & SONS

New York · Chichester · Brisbane · Toronto · Singapore

Please note that the masculine pronoun has been used throughout this book as a matter of convenience only. It is recognized that both sexes can and do function as construction workers.

Library of Congress Cataloging in Publication Data:

Clyde, James E., 1901–
Construction foreman's job guide.

(Wiley series of practical construction guides, ISSN 0271-6011)
"A Wiley-Interscience publication."
Bibliography: p.
1. Building—Handbooks, manuals, etc. 2. Construction industry—Management—Handbooks, manuals, etc. I. Title II. Series

TH145.C55 1987   624'.068   86-29000
ISBN 0-471-81660-4

Printed in the United States of America

10 9 8 7 6 5 4 3 2 1

To my beloved wife
Ruth
this book is dedicated

Series Preface

The construction industry in the United States and other advanced nations continues to grow at a phenomenal rate. In the United States alone construction in the near future will exceed ninety billion dollars a year. With the population explosion and continued demand for new building of all kinds, the need will be for more professional practitioners.

In the past, before science and technology seriously affected the concepts, approaches, methods, and financing of structures, most practitioners developed their know-how by direct experience in the field. Now that the construction industry has become more complex there is a clear need for a more professional approach to new tools for learning and practice.

This series is intended to provide the construction practitioner with up-to-date guides which cover theory, design, and practice to help him approach his problems with more confidence. These books should be useful to all people working in construction: engineers, architects, specification experts, materials and equipment manufacturers, project superintendents, and all who contribute to the construction or engineering firm's success.

Although these books will offer a fuller explanation of the practical problems which face the construction industry, they will also serve the professional educator and student.

M. D. MORRIS, P. E.

Preface

This book is directed mainly to the young man (or woman) who is in good health, is physically strong, likes the out-of-doors, and is trying to decide on a career. He or she does not have to be a technical graduate, but it would help. The book can also be of value to a foreman who wants to broaden his "experience."

To assist you to broaden your knowledge of construction methods and acquire "experience" much faster than on the job, I have listed alphabetically many of the conditions and problems you would expect to encounter on a wide range of civil engineering projects.

The solutions to problems and the methods suggested are based on my over 50 years of experience as an engineer, superintendent, and contractor. They are not necessarily the only solutions, but they worked for me and can form a basis for you to develop your own background in planning, layout, procedures in handling labor, equipment, and material, and keeping records.

JAMES E. CLYDE

Lake Havasu City, Arizona
March 1987

Acknowledgments

My thanks go to the following for permission to use photographs, charts, and other data from projects under their control: Onondaga County (NY) Department of Drainage and Sanitation; O'Brien & Gere, Engineers, Inc., Syracuse, NY; Barton & Loguidice, P. C., Consulting Engineers, Liverpool, NY; John Fura & Son, Inc., Contractors, Verona, NY; Gregory Williams and William McDonnell of the Volpe Construction Co., Inc., Malden, MA; and R. Robertson of McGuire & Bennett, Co.

J. E. C.

Contents *

* Numbers appearing in parentheses are reference numbers for terms that appear in the Construction Specifications Institute (CSI) Index.

4   *60*

1

1.1. Basic Ideas

You may be thinking you would like to become a construction foreman. It is hard work, but a healthy and rewarding vocation. It is not for those who are physically or mentally lazy or for the "9-to-5" crowd. However, for those who are ambitious to get ahead and who will make the effort necessary to acquire the basic knowledge, there are plenty of opportunities for well paying jobs with good chances for advancement. Many of today's successful contractors started as foremen.

No construction project of any size can be successfully and profitably completed without capable and loyal foremen.

As a successful foreman, you must know what to do, know how to do it, and be able to direct labor and equipment for maximum efficiency to get the work completed correctly, on time, and at a profit. Right here it might be well to note that "profit" is not a dirty word. Without profits, there would be few if any worthwhile jobs.

Your most difficult task will be the efficient handling of labor, both skilled and unskilled, who work under your supervision. You must know the capabilities and limitations of each worker and place him where he will work most efficiently. The same is true of each piece of equipment assigned to you. You must know its capabilities and its limitations. Each worker and each piece of equipment will do some jobs better than others, and each must be placed accordingly. Most labor will work best when doing something they like to do, and most will work better with some persons than with others. Some operators will handle one piece of equipment better than another. You must learn these combinations and utilize them to the best advantage.

Most foremen on civil engineering projects come from one of two sources, although there are other routes.

1. The foremen of the specialized crafts, such as pipe fitters, electricians, carpenters, masons, plumbers, and so on, usually come up through the trade and will have been "journeymen" or "masters." They will supervise workers of their own craft. They may be union members and, as such, may retain that membership as foremen. If they have been strong, active union members, they may find it difficult to adjust to a neutral or management-oriented position, but they must if they aspire to continue up the supervisory ladder. The further up the ladder they go, the more management oriented they must become.
2. The other foremen come from the ranks of graduates of high schools, vocational or community schools, and technical or engineering colleges. They will supervise earthwork (excavation, embankment, trenching), sewer and water pipelines, form erection, placing reinforcing bars, concrete pouring, road, street, and airport construction, dams, tunnels, etc. A large percentage of these foremen start with the smaller, usually nonunion, contractors and work at a wide range of duties rather than with one craft.

A foreman does not have to be a top-flight craftsman. It is much more important that he can handle personnel and that he enjoy planning for and directing others. He must know the work well enough to judge if it is being done correctly, according to the plans, and if the production is satisfactory. The foreman on civil engineering construction should know how to do many different types of work, and to get that knowledge it is recommended that he work in a construction crew for at least two years. Most contractors are interested in training foremen in their methods of doing the work and will give a promising young worker a chance to work in different crews on a wide variety of jobs.

Two years may seem like a long time to a young person, but if you plan to make construction your life's work and you hope to advance, the wider the base of your experience, the better. While a member of a crew, observe how things are done. If a mistake is made, and we all make them, try to determine what caused it and how it was corrected. If your foreman is good at his job, study his methods and try to remember them. If he is not effective, try to determine how he could have acted to avoid his problems.

When, as a foreman, you are handed an assignment, be sure you fully understand what is to be done, what materials and equipment will be needed, and the makeup of your crew. You must know the capabilities of each worker, what he does best, and which workers work best together. If you have had the same crew before, you should pretty well know how to place them. If the crew is new to you, you will have to talk to each member and try to determine where best to assign him. Try to put each worker where you think he will function best. With a new crew, you may have to observe each for a while and determine if he would work better at another job. You will not be able to place each worker where he would like to be at all times, and you must place him where you think he will function best. Once you have assigned a worker, require him to stay there until you determine if he might work better some place else. You are the boss, and he will stay where you want him. Of course, you must know the work well enough to know how it should be done. Some workers are chronic complainers and will gripe no matter where they are assigned but will do the work well. Others complain to test their foreman and try to show they are smarter than he is. That is why you must know how best to do the job. On occasion, a worker will suggest a better method of doing a certain piece of work, and if he does, adapt it and give him credit. Once the workers have been assigned and work is proceeding, study what is to be accomplished that day. Are there any obstacles in the way that could delay the work? And if so, can they be removed ahead of the work? If you must work around or through an obstacle, plan how to do it beforehand. If equipment will be needed, will it be on hand and ready to operate? If materials are needed, will they be on hand at points of easy and quick access? Do members of your crew know how to use and place the materials? If not, will skilled help such as equipment operators, carpenters, or a powderman, if blasting is necessary, be available? If you are not sure everything and everybody will be available when needed, confer with your superior and arrange to have them as needed so as not to slow the progress. Delay of completion of a small part of the project can delay completion of the entire job. You should know how the work of your crew is coordinated with and affects the progress of other crews on the project.

The tools needed for the work should be sound and sharp. Power tools should have safety guards in place and working properly. Electric tools should be grounded and the cables placed so as not to trip anybody or be run over by equipment. Gasoline-motor-powered equipment will be heavier than electric, so the workers should be strong enough to handle them and keep them under control. All equipment should be serviced at regular intervals as required.

To coordinate all these problems, you should get your supervisor to advise you at noon today on what your crew will be doing tomorrow, unless yours is an ongoing job. In either case, you should go over in your mind today what the crew will be doing tomorrow, even for an ongoing job. Plan what changes you would have to make if inclement weather were to develop. Are your workers inside where weather is not a factor? If not, will protective clothing be provided? Will runoff rainwater affect the operations? Will pumps be needed, or will a little trenching divert the water? If the work is stopped, is there some other job the crew can move to? Will the weather affect the operation of any equipment used by or with your crew, and if so, what can be done to keep it operating as necessary? Moving equipment may need run-of-bank gravel to stabilize muddy areas. Heavy equipment such as a backhoe or crane may need a timber mat.

Just what labor, equipment, and materials will be needed will depend on what is to be accomplished, under what conditions, and your employer's method of doing the job. As each contractor develops his business, he accumulates certain units of equipment and builds his methods around that equipment. The methods of various contractors for doing a certain piece of work will be similar but not necessarily the same. As a foreman, you will be expected to do the job by his methods, so you can use his equipment. The various items of work are similar enough that if you know one method, you can quickly adjust to the contractor's method. By planning ahead and visualizing each operation, you can anticipate problems and have time to check the contents of this book for advice on most of the problems you will encounter. The methods may not be exactly the same as your contractor's, but they are successful ones and will help you until you learn his way.

Your advance planning must include care for any materials that will require protection from the weather. Will some items have to be stored off the ground? Will tarpaulins or plastic sheets be needed for cover? Also, will there be a toolbox or shed available for storing small tools and equipment at night, or will they have to be returned to a warehouse?

Many workers get upset if they are not paid what was expected and at the scheduled time. Most large projects will have a timekeeper who will check each crew and piece of equipment several times each day and record the time worked. It is recommended that each foreman keep his own record of time, even if there is a timekeeper. You must know each worker's name—be sure you spell it correctly—and his classification and rate of pay. Each day—do not wait until tomorrow—record the hours for regular and overtime.

If the project is a union job, the wages and benefits will be set by agreement between the union and project management, and the foreman will have no say on the matter. However, on nonunion jobs, the foreman can have considerable influence on the rates. On such jobs, where a worker is doing unusually good work the foreman can quite often get his worker a higher rate or have him reclassified to a higher rate. Most contractors will pay better wages for better work where they can deal with individuals.

You should call each worker by his correct name. You should be friendly, but not a buddy, with each. Have lunch with them on the job site, but do not include them in your after-hours social affairs, except maybe at a Christmas party or other special occasion. This can be done without being snobbish and without losing the respect of your crew.

When possible, see that the workers have a clean, shady place to eat their lunch. Clear, cool drinking water must be available at all times, and there should be throw-away paper cups.

You should insist that your workers be properly clothed with heavy work shoes and a hard hat. The rest of the clothing is not too important if the work is indoors. However, some foremen require work pants instead of shorts in the summer as protection from bruises, scratches, or cuts.

Most large projects will have trained first-aid personnel that can treat minor accidents and know when to call in professional help. It is recommended that all foremen complete the Red Cross first-aid course and have a first-aid kit on hand. The contractor's insurance company will provide accident report forms, and you should see that they are filled out promptly, as they can be valuable in case of a claim at a later date. The more details, the better.

When a worker does a good job he should be complimented and any such compliment should be given in the presence of the other workmen. Should it be necessary to criticize a worker, do so in private. In either case, what you say should be the truth, short and to the point. Talk in a moderate voice: no flowery praises, no vulgar criticism.

If a worker puts in full time and does his job correctly, his personal life off the job is no concern of yours. However, no worker can do his best on the job if he is worrying about his personal problems. If you observe one of your crew who has been a consistently good producer, but who has suddenly seemed to slow down or lose interest, you can discreetly ask him if he has a problem and if he wants to talk about it. Sometimes you, or better yet the employer, can, within reason, offer temporary assistance that will help the worker overcome his problem. If a worker knows his boss is interested in his welfare, he usually responds by doing better himself.

The superintendent will expect his foremen to function with a minimum of supervision. The instructions may be verbal or in a memo, and there may be sketches and/or blueprints. I repeat, each foreman must immediately assure himself that he fully understands what is to be done, what crew, equipment, and materials will be available, and how to use them. If he does not fully understand, he must confer with his superior at once. Do not try to bluff it through: correcting mistakes is costly and time consuming. Nobody is expected to know everything about everything, and anyone who asks questions will be respected for being sure before starting.

Some contractors employ "working foremen" who work right along with the rest of the crew and more or less pace the progress. Such projects will usually have a "general foreman" for overall supervision. Some firms will start you as a working foreman just to see how you handle yourself, and if you work out, they will advance you to full foreman after a few weeks. One problem of a working foreman is that he may become so interested in what he is doing that he may fail to notice that some of the workers are watching him instead of doing their share of the work.

You will note that this book is generally oriented to civil engineering projects, the so-called heavy construction. However, much of the work described also relates to building construction. The methods for excavation, footings, concrete or masonry walls, structural steel or reinforced concrete framing, forms and falsework, roofing and flashing, and landscaping also apply to building construction.

To summarize, a good foreman knows what he has to do and how to do it efficiently with his employer's labor and equipment. He will, each day, set reasonable goals and motivate his crew to attain them. He will recognize problems that could delay his progress and will plan ahead how to deal with them. He will try to have other work available in case the scheduled operations are disrupted.

2

2.1. Abutments

Abutments are the end supports for bridges and for arches in other structures. They were formerly made of stone masonry with large cut stones laid up with or without mortar. Today they are of poured concrete and may have a stone facing for architectural effect (see Section 17.14).

On bridges, in addition to supporting the span, an abutment must retain the earth fill behind it, the same as a retaining wall, and should have weep holes for drainage (see Section 19.9).

If the span is of concrete, the abutment will be shaped to receive the end of the arch. If the span is structural steel, the abutment will have a steel or cast-iron bearing support (see Section 3.11). Like any support structure, an abutment must rest on a solid subgrade or piling. The plans will indicate which [(see Section 20.61) (Figure 2.1)].

2.2. Access Roads

Access roads can be permanent secondary roads leading to main highways or temporary roads for access to a construction site. The permanent roads will be under municipal control and subject to their requirements. Roads to a construction site are usually built by the contractor involved and maintained by him. Sometimes such roads later become permanent, being taken over by the municipality after construction is completed, or the area may have to be restored.

If you are responsible for the access roads, some of your problems on permanent roads will be traffic flow, flagmen, spills, daily cleanup, barricades,

FIGURE 2.1. The abutment of a two-span highway bridge. Actually, it is one-half of a bridge; the other half will be built after the old bridge in the background is removed.

Note the crushed stone placed behind the abutment wall for drainage. Steel sheet piling is holding the backfill until the other half of the bridge is completed. When completed, the structure will be a four-lane highway bridge.

signs, and warning lights. On construction roads your problems can be maintenance of the road surface, drainage, assisting stuck vehicles, and traffic flow. Trucks and other vehicles coming onto a paved highway from a construction site will deposit mud and gravel, which can cause traffic problems. Small quantities can be removed by square-point shovels, but larger amounts will need a road-grader blade. Either one will leave a film of dirt, which will be slippery when wet and cause dust when dry. The dust can be controlled by spreading calcium powder.

In some areas, the use of calcium is forbidden by environmental regulations, in which case sprinkler trucks are used. If there is a heavy accumulation of dirt, powered sweepers may be used.

The amount of maintenance undertaken will depend on the policies of the contractor. Some contractors will endeavor to keep things in good order, while others will do as little as possible and move only when a municipality orders them to or the vehicles get stuck and hold up the job. You will have to be governed by what your instructions are.

Quite often, construction roads are built with too little base, and the trucks break through the surface and get stuck. Usually, they will develop ruts which get deeper until finally the trucks get stuck. A stuck truck can be pulled or pushed free by another truck unless the axle is on the ground, in which case a bulldozer will be needed to push it free.

Of course, it is more economical to construct a proper base at the start, but many contractors will try to get by doing as little as possible.

Permanent access roads will be built according to the owner's specifications with a properly prepared subgrade, base course, and wearing surface (see Section 19.15).

Temporary access roads on a construction site will be developed as the work progresses, and the amount of work done on them will depend on the soil and weather conditions. In high and dry areas, very little work will be done beyond driving vehicles over a designated route. If problems develop due to soft areas or wet weather, crushed stone or run-of-bank gravel will be spread where needed. In low swampy areas, a whole length of road may have to be prepared before work starts. One method is to lay a mat of brush and small trees and cover it with crushed stone or gravel. Another is to lay a sheet of heavy fiber or plastic and cover that with stone or gravel. A good method is to lay about 3 in. of well graded run-of-bank gravel with crushed stone over that to whatever depth is required. The gravel is less likely to punch holes in the plastic sheet, and the crushed stone will bond together better under traffic.

Low swampy areas should be drained as much as possible (see Section 7.25).

2.3. Accidents

Unfortunately, accidents do happen, even on the best-regulated jobs. You must be constantly on the alert for conditions that might lead to an accident. Most contractors will have some kind of a first-aid kit on the job, and the larger ones will have a properly trained first-aid man. Many contractors hire a trained specialist who will be both a "safety" man and first-aider. All accidents, including minor ones, should be reported as soon as they happen, as complications can set in later. If a first-aid man is available, he should examine the injured person and treat him as necessary.

Minor cuts and bruises can be treated by cleaning the cut and applying a protective bandage. A bruise may or may not require a bandage. When a slightly injured worker returns to the job, he should be observed for a while, as some persons experience shock even from minor injuries.

When a worker is injured by a fall or struck by some object and no visible injury is observed, he should not be moved until a qualified person examines him. He may have broken bones or an internal injury that would be aggravated by movement.

The contractor will have a report form, usually supplied by his insurance company, which should be filled out in detail. That task often falls on the foreman. All questions should be answered in as much detail as possible. The report may be the basis of a claim at a later date.

The necessity for constant observance for conditions that might lead to an accident cannot be overemphasized. Many otherwise capable foremen have been discharged because their crews experienced too many accidents.

There are many inexpensive booklets on the market which deal with first aid and treatment of injured persons. It is recommended that each foreman have such a booklet and study it. The Red Cross and many fire departments have courses you may attend.

Some of the causes of accidents are

1. Nails in boards on the ground
2. Objects falling into trenches or from overhead
3. Collapse of unstable trench banks
4. No traffic control of moving vehicles
5. No alarms on backing-up equipment
6. Improper scaffolding, ladders, elevators
7. Old or weak cable or rope
8. Insecure forms or falsework (concrete)
9. Equipment touching overhead bare wires (cranes, backhoes, well-point jet pipes, etc.)
10. Insufficient warning for blasting or improper cover
11. Slippery surfaces
12. Ungrounded cable for power tools
13. Lack of barricades and warning lights
14. Insufficient lighting for night jobs and in tunnels
15. Lack of equipment guards
16. Operator unfamiliar with equipment
17. Lack of hard hats, protective glasses, and gloves

A contractor cannot afford to keep an employee who is accident prone.

2.4. Aeration

Aeration tanks at a sewage disposal plant inject air into the sewage to increase the action of the aerobic bacteria. The air is usually introduced in such a manner as to cause the sewage mass to rotate and expose more of the sewage to the sunlight and open air. There are several methods of injecting the air. One method is to have porous filter plates placed along one side of the tank. Another is to place small air jets every few inches along a horizontal pipe, which may be supported along the bottom of the tank or suspended from the surface (see Figure 2.2). Either system must be airtight, with the air being released only through the filter plates or the air jets.

After the system has been installed, the tank should be filled with clean water to a level about 12 in. above the filter plates or the air jets, and the air should be turned on at low pressure. The pattern of air bubbles should be uniform over the full length of the tank. If the agitation of bubbles is greater in one spot than another, the system must be adjusted to make them uniform. Check for leaks at the pipe joints or around the edges of the filter plates. If the water is clean, you will be able to see where the leaks are and mark them for adjustment.

Another method of aeration is the oxidation pond, whereby sewage is pumped into ponds of large surface area and shallow depth. The sewage is allowed to stay in these ponds for a considerable time, exposed to the open air and sunlight. Such ponds may be lined with watertight sheeting or a layer of clay. If an embankment is involved, care must be taken to see that the earth is well compacted.

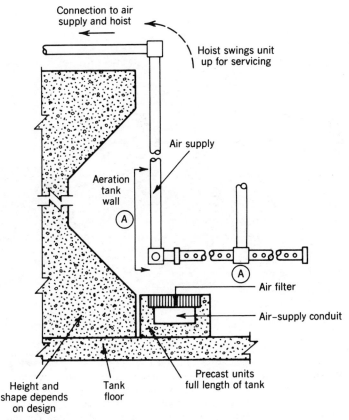

FIGURE 2.2. Two different systems for supplying air. The pipes are serviced by hoisting them above the water. The present box requires the emptying of the tank.

2.5. Aggregate

Aggregate includes run-of-bank gravel, sand, crushed stone or gravel, graded sand, and slag. It is used as a base material for roads, streets, runways, parking lots, etc. It is also used for making Portland cement or bituminous concrete, mortar, stucco, plaster, and cement grout.

The plans and specifications will indicate the kinds and gradations of the aggregate for the various parts of the work. On jobs of any size, laboratory tests will be required, and the foreman can usually assume that the engineer or inspector will have arranged for tests and that aggregate accepted on the site will have been tested and approved.

Aggregate stored on the site should be kept free of mud and dust. Care should be taken to see that equipment picking up the aggregate does not also pick up some earth, especially when the pile gets low.

Sand is called "fine aggregate," and all other, "coarse aggregate" (see Figure 2.3).

FIGURE 2.3. A stockpile of "run-of-crusher" stone, which will be used as lining under and around a 24-in. corrugated metal pipe for a storm sewer.

When large pieces of rock or stone are placed in a crusher, they are broken down into pieces ranging from those just small enough to fall between the jaws of the crusher down to small chips and dust. The jaws can be set to control the maximum size. When used as it comes from the crusher, it is called run-of-crusher and is a mixture of all sizes. The mixture can be run through a series of screens, and the different sizes can be sorted out. The mix in the photo has a maximum size of 2 in.

2.6. *Air Compressors*

Most civil engineering projects will have one or more portable air compressors on the job to operate power drills, jackhammers, pile-driver hammers, spraying equipment, and numerous types of power tools. Except for the pile driver, one compressor will handle several tools at once, depending on its size. If several tools are to be operated at one time, it is recommended that the compressor be set up as near to the center of operations as possible. This will keep the air hoses shorter and less likely to interfere with other operations. Each air hose will have a lubricator in the line to supply oil to the air tool. Be sure the lubricator is always full of the proper grade of oil.

The air compressor will vibrate considerably, so it should be set on firm ground or on planks to keep the wheels from sinking into the ground. Plenty of fuel should be on hand to avoid delay waiting for a supply.

Air compressors are also part of the permanent equipment in many plants. If the units are large, they will usually be housed in a separate room with soundproof walls and large louvers in the outside wall to supply air.

If you are involved in setting up a permanent compressor, see that all anchor bolts are in place and the nuts are tightened. Study the plans for vibration-dampening devices in the supply lines. The lines may be of sheet metal or plastic and may be supported on the floor or hung from the ceiling. The setting of the lines will be up to the sheet-metal workers, but you may be assigned to assist.

2.7. *Air Conditioning*

Air conditioning can be supplied in three ways: by individual units through an exterior wall, usually under a window; by individual stand-up units placed throughout the building; or by central units with metal or plastic ducts to distribute the air. The individual units will serve one medium-sized room, the stand-up unit will service one good-sized room of 1000 ft^2 or more, and central units will serve the whole structure, with the ducts serving for both the air conditioning and heating.

The large central units will be installed by the heating–ventilating–air-conditioning (H–V–A/C) contractor. Your crew may assist in placing the units and distributing the ductwork through the building.

2.8. *Air Entraining*

Air-entrained concrete is used to resist frost action and where salt, such as in seawater or as used on pavements to reduce snow and ice, comes in contact with the concrete.

It is also used in concrete for exterior walls, parapets, walks and steps, and retaining walls that come in contact with the soil or are exposed to extreme weather conditions.

The air-entraining agent will have been added as the concrete is mixed, so you will have no special duty in handling it. Just place the concrete as usual. A laboratory employee may come on site to check the mix.

2.9. Alarm Systems

Alarm systems are used to detect intruders and to warn the operator of a malfunctioning of equipment anywhere in the system. The systems will be indicated on the plans and in the specifications. They will be installed and put in operation by the electrical contractor (see Section 4.47).

2.10. Alignment

Alignment has to do with placing and keeping the various parts of a project in their proper position. Walls must be straight and plumb, water and sewer lines must be at the proper line and grade, roads and streets must be on proper line and grade, columns and piling must be in the proper position and plumb, and so on.

The various items are placed on line and grade by various means. Building walls and columns are controlled by batter boards, and water and sewer lines are controlled by grade stakes with grade line and grade pole or by laser beam. Each of these methods is explained under its own heading (Figures 2.4–2.8).

Railroads

Railroads are laid from centerline grade stakes. The top of the stake will have an elevation relating to the base of the rail. A tack in the top of the stake will give the centerline of the track between the rails. The profile of rail tracks is set at the base of the rail, so, knowing the elevation of the base of the rail and the elevation of the top of the stake, the distance X can be calculated for setting the track. (See Figure 2.8)

Bridges

Most bridges are laid out from centerline stakes for lines and nearby bench marks (BMs) for elevations. Large bridges over wide rivers or lakes are laid out by surveyor's instruments, which will be in constant use until all foundations, anchor bolts, and bearing plates are set.

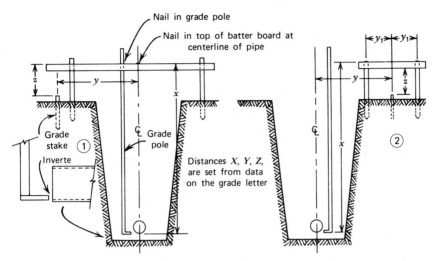

FIGURE 2.4. Sewer pipe alignment. (1) Batter board spans the trench. Top of board set from grade stake to an even-foot distance above the invert of the pipe. Nail in top of the board is centerline of the pipe and set from nail in top of grade stake. (2) Batter board set on one side of the trench. Mason's line from one batter board to the next gives centerline and grade at any point between the batter boards. Nail in grade pole gives distance from line to invert of pipe. Plumb bob at line gives the centerline of the pipe.

Tunnels

Most tunnels are now laid out by use of the laser beam, although surveyor's instruments are still used.

Tanks

Standpipes or tanks set on the ground will have a circular concrete wall foundation laid out from a centerline stake with a tack and elevation on the top.

Canals

Canals and large ditches will be laid out from centerline and slope stakes.

Levees

Levees are staked by slope stakes similar to those used in constructing highway embankments.

It is your responsibility to see that all work is constructed as indicated by the layout stakes. On many projects, the surveyor sets general control stakes only, and the contractor's people set whatever work stakes they want

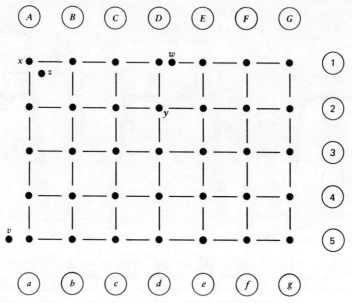

FIGURE 2.5. Piling stakes. The surveyor sets stakes at points A, B, C, and so on, a, b, c, and so on, and 1, 2, 3, and so on. If the distances are small, a mason's line (or wire) can be stretched from A to a and the distances A–1, 1–2, 2–3, and so forth can be hand measured. On longer distances, a survey instrument is set on A, sighted on a, and the distances A–1, 1–2, 2–3, and so on are measured. A stake is then set for the center of the pile. Piles X and Y are on intersections, and piles W, Z, and V are on offsets and must be located from the other lines. The plans have a sketch similar to the one shown to show the distances required.

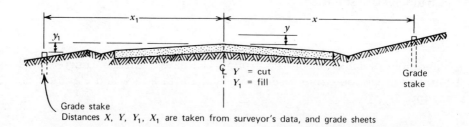

Distances X, Y, Y_1, X_1 are taken from surveyor's data, and grade sheets

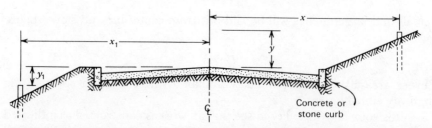

FIGURE 2.6. Highway stakes. The distances X, Y, $Y1$, and $X1$ are taken from the surveyor's data and the grade sheets.

16

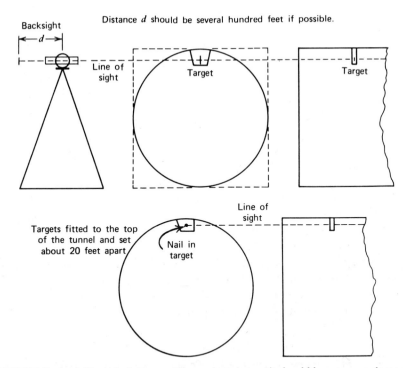

FIGURE 2.7. (top) Tunnel alignment. The surveyor's transit should be set up as close to the entrance to the tunnel as possible. Make a backsight that puts the transit in line and grade. Project the line into the tunnel, making a check on the backsight as often as necessary. A target shaped to fit the top of the tunnel can be placed on the line of sight and the distance off line and off grade can be checked. Corrections, if any, can then be made to bring the tunnel back on line or grade. Targets set at the top of the tunnel are set about 20 ft apart. (bottom) For tunnels on which the line and grade are not so important, the work can be kept reasonably close by setting a target at the entrance and others (at 20 ft) as the work progresses. The line and grade can then be checked by sighting along the top of the nails.

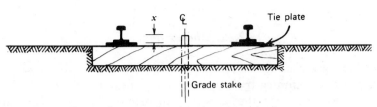

X = cut from top of grade stake to base of rail

FIGURE 2.8. Railroad stakes. X is the calculated distance from the top of the grade stake to the base of the rail.

from the controls. You must be fully informed as to just what the various stakes represent and how they are used. If you are unsure, get the correct information before construction starts.

Some pipeline contractors use a laser beam for setting their pipes. In such cases, the surveyor places offset stakes indicating line and grade at each manhole or other intermediate points. The use of a laser beam can be dangerous to the eyes. The placing and operation of the equipment should be done only by persons properly trained.

Force mains are usually not set to grade, but follow the contours of the earth's surface at a designated depth. The high points are determined to indicate locations of vacuum release valves, and the low points are determined to indicate locations of blowoff valves.

Water mains under 18 in. in diameter are usually installed following the earth's surface, in the same way as force mains. Pipes 18 in. and over in diameter are usually set to an established grade.

The larger diameter pipes, which are set to grade, have sections of the pipes cast with angled ends for horizontal and/or vertical curves. In such cases, the individual pipe sections are numbered and are placed in accordance with a laying chart furnished with the pipe.

2.11. American Society for Testing Materials (ASTM)

The specifications, when referring to the quality of materials, will refer to the ASTM requirements. The ASTM is an organization that conducts extensive tests on all kinds of construction materials and issues a series of booklets on the results of these tests, with recommendations on the limits of their use. The design engineer will follow these recommendations when designing the structure. The contractor will follow them when ordering materials.

2.12. Anchor Bolts

Anchor bolts are used to hold in place various parts of a structure such as beams, girders, columns, trusses, and some types of exterior covering of walls and to secure in place the bases of permanent machinery and equipment. You must see that the bolts are placed exactly as shown on the plans for both position and height above the foundation. They should be securely wired in place so they can not move while the concrete is being poured, and they should be rechecked as soon as the concrete is poured just in case some did move.

A careful study of Figures 2.9–2.11 will help you in setting the bolts. After the concrete has set, the bolts must be protected until everything

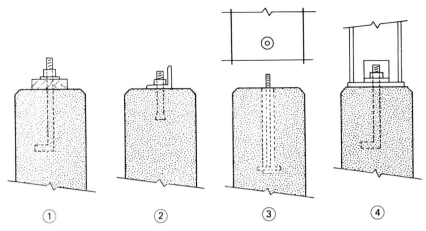

FIGURE 2.9. Anchor bolts. (1) Cast-in-place anchor bolt for wooden sill. (2) Expansion bolt in drilled hole to anchor steel angle. (3) Cast-in-place anchor bolt in a sleeve to allow adjustment to fit holes in equipment base. After bolts are adjusted to fit the equipment, the void is filled with hot lead or grout. (4) Cast-in-place bolt for steel column.

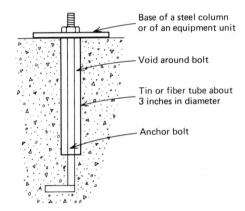

Base of a steel column or of an equipment unit

Void around bolt

Tin or fiber tube about 3 inches in diameter

Anchor bolt

FIGURE2.10. When it is difficult to determine the exact positions for the anchor bolts, they can be installed as shown. After the column base or equipment is on site, the exact position can be measured and the bolts moved so as to conform with the bolt holes. After the bolts have been positioned, the void will be filled with grout before the base is set.

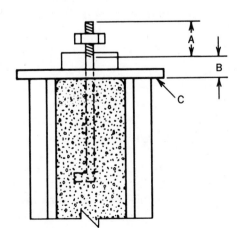

FIGURE 2.11. A, length of thread. B, Distance of thread above concrete. C, One or two inch board, four to six inches wide (depending on size of bolt). After the concrete has been poured, it should be well vibrated around the bolt, and the bolt should be adjusted so that the bottom of the threads will be slightly below the top of the base plate. This will allow the nut to be fully tightened.

19

is tightened in its final place. This can be done by placing the nut on the bolt at about the center of the threads and wrapping them with heavy paper or fiber.

2.13. Anticipate Problems

A forecast of inclement weather should alert you to be prepared for rain, snow, or high winds. You may have to remove snow or ice before starting work. If rain, you may need extra pumps and/or have to construct ditches to carry off the storm water and keep it out of the excavation.

If some equipment has not been operating up to par, you might want to request a replacement to avoid a slowdown in production.

If your work area tends to get muddy after a rain, there should be a stockpile of gravel available to provide traction for vehicles.

If your equipment has not been serviced regularly, you should talk to your superior, as lack of proper servicing can result in damage to machines. Also, there should be enough fuel on hand to keep your equipment operating all day. Arrange to have all tools sharp at the start of work.

Anticipating and avoiding these problems will help keep production up.

2.14. Arbitration

Sometimes a disagreement arises between the contractor and the engineer or architect over the quantity and/or quality of parts of the work. If they cannot arrive at a settlement of the problem, they may submit the case to arbitration by others. This is usually done through the American Arbitration Association. Some specifications require such action, and it is binding on both parties. The foreman may be called to testify as to what he knows about the problem. In such cases, tell the truth as you know it, whether it happens to hurt the contractor or not. Do not volunteer any information; only answer the questions directed to you.

2.15. Arches

Arches can be made of reinforced concrete, masonry, structural steel, or laminated wood. Unlike beams and girders where end loads on their supports are applied vertically, end loads from arches are applied at an angle, so care must be taken to follow the plan details.

Reinforced concrete arches are poured into forms supported on false-work. On heavy, long-span arches, the forms will be set slightly above the design elevation to allow for settlement as the forms and falsework are loaded. The amount of settlement will depend on the design of the forms and falsework and the care taken in erecting them. There will be a slight takeup at every point where one timber bears on another. A rule of thumb of 1 in. for every 100 ft of span works rather well. In any case, settlement will have to be controlled by wedges or screw jacks (see Sections 7.1 and 7.22).

The support on which the end of the arch rests is called a "skewback" and will have dowels to match the reinforcing bars in the arch. On long-span arches, there will be many bars, which can make the placing of the concrete difficult, especially in the area of the dowels. This is an area where there is quite often an argument between the inspector and the contractor over how much water can be in the concrete. The wet mix is easier to place, but it also lowers the strength of the concrete. You will have to follow your superior's instructions. A plastic mix can be placed and well vibrated into the corners and around the bars by the use of a small vibrator head.

Pounding on the outside of the forms will help settle the concrete, and there are also vibrators that can be attached to the outside of the forms; however, unless properly used, they can damage the forms.

When pouring concrete into arch forms, you start at the ends and work up the arch evenly on both sides. Some sort of gauge should be set to record the settlement as the forms are loaded, or a surveyor's level should be set up for checking. The placing of the concrete should be the same as for any other structure.

Structural-steel and laminated-wood arches will come to the job site shaped to the proper lines. For short spans, the arch will be in one piece. For longer ones, it will usually come in two sections with a bearing or hinge at the top center. Such arches will usually be erected by steelworkers or maybe carpenters. The ends will be secured by bolts or welding.

2.16. Areas (see Appendix C)

2.17. Argue

Never argue with your crew. To do so only brings ill will, and you lose control.

If there is a disagreement, talk it over calmly. First, be sure you are right in your position and that you are not asking for an impossible task. Explain your position clearly and then insist on your instructions being followed. If a worker thinks he knows a better way, listen to him. He may be right. Do not hesitate to use a suggestion from a member of your crew if it will benefit the work, and be sure to give him credit for the idea.

2.18. Artesian Wells

Artesian wells are caused when the underground hydraulic pressure raises the water above the surface. The height above the surface can be from a few inches to several feet. The boils in a quicksand area are small artesian wells. The pressure can be enough to raise the basement floor if provision is not made to carry the water off. Some floors are designed to take the pressure. When artesian wells are found on a construction site, various methods are used to lower the groundwater elevation (see Section 8.18). If the soil is porous enough to allow free flow, wells may be placed around the area. French drains are used to carry the water to wells for pumping or to a lower elevation where it will flow naturally, or well points may be required. (see Section 24.14).

2.19. Asbestos–Cement Pipe (A–C)

Asbestos–cement pipe is made of Portland-cement mortar mixed with asbestos fibers. The pipe is lightweight and has reasonably good strength, although it is not as strong as metal pipe. It will resist the buildup of deposits on the inside much better than metal pipe.

The pipe is subject to shearing stresses and can break if not well and uniformly bedded. The joints and the gaskets must be well lubricated with the compound furnished. Care must be taken to see that the gasket is not displaced when making a joint. A metal gauge will be furnished to check the gasket by running the gauge around the joint to see that the gasket is in the proper place. A misaligned gasket can cause a leak. A common practice is to align the spigot in the collar and ram it home by use of a crowbar at the other end, using a block of 2 × 4 to protect the end of the pipe (see Figure 2.12).

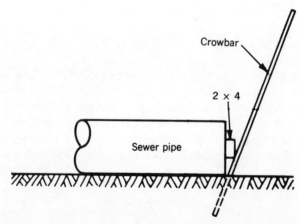

Crowbar

2 × 4

Sewer pipe

FIGURE 2.12. One method of forcing a pipe joint tight without damage to the pipe.

Because it is expensive to repair leaks, all pipe should be carefully handled and installed, especially A–C. Do not drop the pipe into the trench on its end, as the thin spigot can be easily damaged. If the trench is shallow, the pipe can be handed down; if not, it should be lowered with a rope.

2.20. As-Built Plans

These are plans which show how the work was finally built, including changes in the structure or alignment of pipes, cables, or other underground items such as valves, thrust blocks, and so on. The work of measuring up the changes is usually done by the surveyors; however, the foreman may do so, especially for items he will be covering up.

These plans are used to locate parts of the work at some future date when repairs or changes are required.

2.21. Asphalt

(See Section 3.18).

3

3.1. Backfill

Backfill is the excavated earth that is put back into the excavation after the structure has been installed, such as around a building foundation or in a trench after a pipe has been installed. In most cases, the specifications will require that the backfill be compacted to a specified density, usually 95%. The fill will be placed in layers 8–12 in. in depth, with each layer compacted before the next layer is placed. The compacting is done with various kinds of equipment (see Section 4.32). When backfilling around a foundation wall, care must be taken not to put enough pressure on the wall to move or topple it. Most specifications will require that the walls be braced or the backfilling be delayed until the first floor has been installed. Usually, selected material is required for backfill: no large stones, organic material, or debris. If there is an inspector on the job, he may require an inspection of each layer before the next layer is placed. If there is a footing drain around the foundation, care must be taken not to disturb or damage it as the fill is placed. When backfilling a trench, as pipe is being laid, the first foot should be material selected to act as a cushion to protect from stones in the backfill or which could fall in from the surface. Trenches that are within a street or roadway will require complete compaction for the full depth to avoid settlement, which could disrupt traffic. Coming back to repair settlement is costly. It is much more economical to do a thorough job in the first place, and it is much better public relations. In winter weather, frozen chunks should not be placed in the backfill, as they will cause settlement when they thaw out (see Figure 3.1).

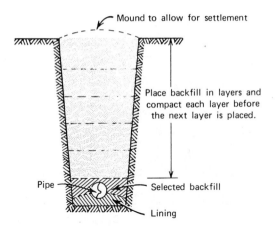

Mound to allow for settlement

Place backfill in layers and compact each layer before the next layer is placed.

Pipe

Selected backfill

Lining

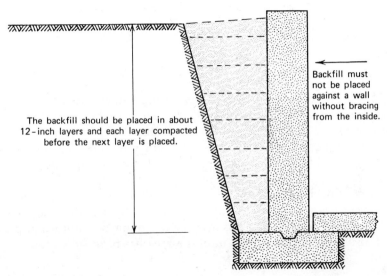

The backfill should be placed in about 12-inch layers and each layer compacted before the next layer is placed.

Backfill must not be placed against a wall without bracing from the inside.

FIGURE 3.1. (top) Backfill. Most plans require that the pipe be bedded in lining and selected fine backfill be placed around and over the pipe to a depth of at least 1 ft. The backfill above that is usually the material excavated from the trench. If the excavated material is not suitable, special backfill is used and paid for under a "special backfill" item in the contract. (bottom) Backfill should be placed in layers, and each layer should be compacted before the next layer is placed. No layer should exceed 12 in. Backfill must not be placed against a wall without bracing the wall from the inside.

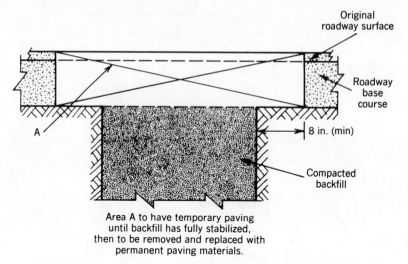

Area A to have temporary paving
until backfill has fully stabilized,
then to be removed and replaced with
permanent paving materials.

FIGURE 3.2. A sequence for backfilling a trench under a traveled way.

Some specifications will require that temporary paving be placed for a specified time or until settlement is complete. Usually, the temporary and final paving will overlap the trench by about a foot (see Figure 3.2).

Backfilling around structures will usually require that the fill at the wall be slightly higher than the general area, to provide drainage away from the structure. Most slopes are set at 6 in. in 10 ft. When backfilling around an area for final landscaping, the contours will be laid out to provide proper drainage. In bringing the soil up to grade, remember that most jobs will require 4 in. of topsoil, and this must be taken into account when grading. The topsoil will not be spread until the area has been fully graded.

3.2. Barber–Greene Spreaders

The Barber–Greene is one of several makes of spreaders for bituminous concrete. They are all similar in that they spread the hot mix to preset widths and depths (see Section 3.18).

3.3. Barco Compactors

The Barco compactor is powered by a gasoline motor that causes the unit to jump about 2 ft in the air and fall back to compact the earth with a steel baseplate. The unit weighs about 125 lb, is about 4 ft high, and has handles similar to bicycle handles for control. As the unit hits the ground, the motor

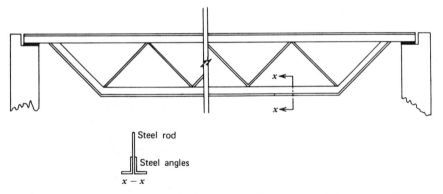

Steel rod

Steel angles

$x - x$

FIGURE 3.3. Bar or open-web joint. The makeup of the open-web joint varies from one manufacturer to another. All, however, are similar to the one shown in the figure.

FIGURE 3.4. Bar joists for the roof of a new store in a shopping center. Note the two different span lengths. Such joists come banded together as shown.

FIGURE 3.5. Bar joists in place for the roof structure. Note longitudinal bars, which keep the joist vertical.

FIGURE 3.6. Stacks of metal forms which will be placed on the bar joists to form the roof slab. Note that in Fig. 3.5 some of the sheets have been placed on the roof ready to set on the joists. These forms stay in place as 3 or 4 in. of lightweight concrete is poured on them. Sometimes a very light gauge wire mesh is placed in the concrete.

fires and forces it to jump, it falls back to earth, and the process is repeated. The unit will keep jumping as long as the motor switch is on and the unit is kept in an upright position. Needless to say, it requires a husky operator.

3.4. Bar Joists

Bar joists or open-web joists are lightweight trusses used for floors and ceilings where strength but light weight is required. They are used mostly in commercial or light manufacturing buildings. The web members will be steel bars with the top and bottom members being small angles, steel bars, or pipe. When the loads are light, the span can be considerable. Their light weight allows fast installation (see Figures 3.3–3.6).

3.5. Base Courses (Roads)

The base course of a road or street is the material placed on the subgrade to give strength for supporting the traffic. There may be one or several layers to provide the depth required to support the type of traffic expected. The material may be Portland cement or bituminous concrete, crushed stone, run-of-bank gravel, or a combination of these. The plans will show the kinds of material and the depths.

Each layer will be compacted by rolling or vibrating, or both, to the density required before the next layer is placed. The engineer may require tests to check the density (see Section 19.17).

Figure 3.7 shows base courses for a roadway. There can be one or several courses and the aggregate in each may vary.

3.6. Base Lines

Base lines are survey lines established before construction starts and are used as reference points from which construction stakes are set for controlling

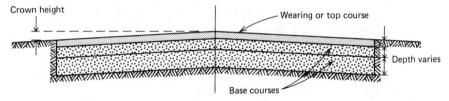

Crown height — Wearing or top course — Depth varies — Base courses

FIGURE 3.7. Layers of a roadway.

the horizontal and vertical alignment of the various parts of the project (see Section 20.45). Foremen are seldom involved with base line stakes, but if you are, you should know that they are often set months and even years before construction starts and can be knocked off line or otherwise damaged by equipment operating in the area. They can also be moved by frost action in the ground. They should be carefully examined, and if there is any question of their accuracy, you should confer with your superior.

3.7. Batching Plants

Batching plants produce either Portland cement or bituminous concrete. Large projects may have an on-site plant, but most jobs purchase the concrete from commercial plants, which produce a mix to meet state and federal requirements, delivering the mix to the site in large dump trucks for bituminous concrete and in $5-15$ yd^3 mixer trucks for Portland cement concrete. In either case, the truck driver will have a batch ticket, giving the quality and quantity of the load. You should collect these tickets as each load is delivered, as they are proof of delivery, and turn them in to your superior at the end of the day.

3.8. Batter Boards

Batter boards are used to control the alignment and elevation of the various parts of a structure during its erection.

For buildings, the location of the principal corners are set by surveyors, who place a stake with a tack on the top to indicate the exact location of the corner. The top of the stake will have an elevation with reference to some part of the structure, such as the top of a footing or the top of a finished floor. The surveyor will furnish information on how far, in feet and inches (or feet and tenths of inches), the top of the stake is above or below the footing or floor. With this information, the batter boards can be set at even feet above (fill) or below (cut) the footing or floor.

Referring to Figures 3.8 and 3.9, to set the horizontal batter boards when they are within 4 ft of the corner stake, use a 4 ft level, and if they are over 4 ft, use a builder's or surveyor's level. Assuming the top of the building foundation is at elevation 258 ft and you want to use an 8 ft grade pole, then the top of the batter board would be at 258 plus 8 equals 266 ft. Assuming the top of the corner stake to be at elevation 263.48 ft, then 266.00 minus 263.48 equals 2.52 ft, so you must set the batter board 2.52 ft, or 2 ft 6¼ in. above the top of the stake.

With the batter boards set at all four corners, hold the mason's cord on the tack on top of the corner stake at B in Figure 3.9 and pull it taut over the batter board at A. Hold a plumb bob over the tack in the stake at A, slide the cord along the top of the board until it just touches the plumb bob string, mark the board where the cord crosses it, and drive a nail on the mark, leaving the nail about 1 in. above the board. Now reverse the procedure and set a nail on the board at B. A mason's cord pulled taut between the two nails will be on the foundation line. Follow the same procedure all around the building.

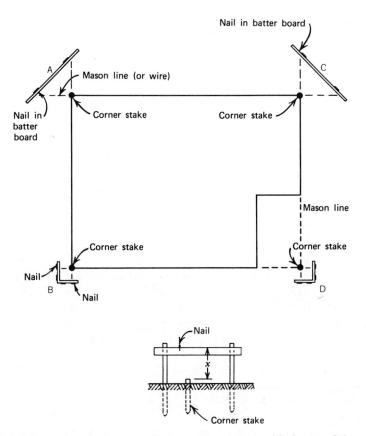

FIGURE 3.8. Batter-boards. Corner stakes are set by surveyor with the top of the stake set at a known elevation. A nail in the top of the corner stake gives the exact corner of the structure. The surveyor sets the corner stakes, and a workman, usually a carpenter, sets the batter-boards to a known elevation and set the nail to indicate the outside edge of a wall or footing. Distance X will set the top of the batter-board at an even-foot elevation or an even foot above the footing.

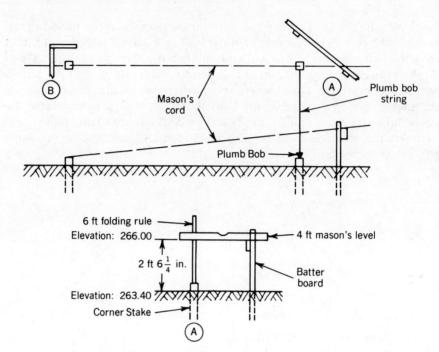

FIGURE 3.9. A method of setting batter-boards.

3.9. Beams (Structural)

A beam is a structural member, such as a floor joist, bridge decking, or a lintel over a door or a window in a wall, that supports a load over an opening. The beam may be of steel, reinforced concrete, or timber, or a combination of these. The ends may rest on a wall, column, or girder. There may be a simple, single span, or the beam may extend over several supports, and the ends may be cantilevered. Most beams will have one fixed end and one free end, for expansion and contraction. The fixed end will be tied into the support in some manner. A steel beam will have a short rod through a hole in the web. For reinforced concrete, the rods will extend over the support. A timber unit will be nailed or bolted to the support.

3.10. Beam Schedules

Concrete

The beam schedule assists the steelworkers in placing the various bars in the position the designer intended. Different engineers have different forms, but basically they will be similar to that in Figure 3.10. The beams and

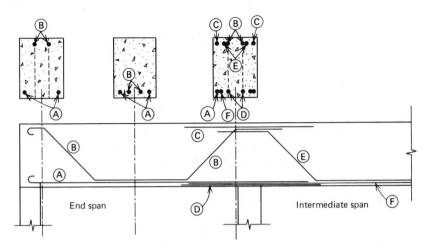

FIGURE 3.10. C and D bars are straight bars over the support. B and E bars overlap the support of intermediate span.

girders will be numbered on the plans, and there will be a list of them with the number of bars in each. Also, there will be a sketch of each type of bar with the dimensions of all bends and lengths. Each type of bar will be identified by letter and/or number, with several like-type bars bundled together.

Structural Steel

A beam schedule for structural steel will be similar to one for concrete, except there will be no sketch, but rather a listing of each beam and girder. The listing will have numbers corresponding to those on the plans, showing the type of unit, such as 12WF85, which means a 12-in-wide flange beam weighing 85 lb/ft. An angle would be listed as 3 × 2 L 6.8, which means an angle with one leg at 3 in. and the other at 2 in. and weighing 6.8 lb/ft.

Some plans will not have a schedule, but will show the size, type, and weight of each beam or girder right on the plans.

3.11. Bearing Plates

A bearing plate for a beam, girder, column, or truss can be a simple steel plate of an elaborate unit with steel rollers or rockers. The fixed end of the span will have a single plate secured to a support surface with two or more anchor bolts, which will also hold the beam or girder. The expansion end will have two plates, one secured to the support surface and the other to the beam or girder. Graphite is often placed between the plates to reduce the friction. For long, heavier spans, much more elaborate units with rollers or rockers that allow several inches of movement are used. The plans will indicate what is required.

The elevation and alignment of the bearing plate is very important, as misalignment or incorrect elevation could cause binding as the beam expands or contracts due to changes in the weather.

Usually, the holes for the anchor bolts will be large enough to allow some movement for precise adjustment. Often, the plate or unit is set on steel wedges for adjustment to the proper elevation, and then nonshrinking grout is packed under the plate after everything is in its correct position.

3.12. Bearings (Machine)

The quality of the bearings in any machine will determine its efficiency and useful life. As most contracts call for a 1-yr guarantee, any problems with the bearings can cost the contractor money.

If you are involved with the installation of permanent equipment, you should check to see if the bearings seem to be properly lubricated. Any excess grease should be wiped off, and if there appears to be a lack of grease, confer with your superior. If possible, turn the equipment by hand. It should turn freely without binding; if not, inform your superior. Sometimes with lightweight machines the uneven tightening of the anchor bolts can warp the frame enough to bind the bearings. The anchor bolts should all be tightened to the same degree (see Section 21.22). If the equipment turns freely by hand, start it up and listen for noises from the bearings. They should produce a low, steady hum with no chattering. Chattering indicates problems and should be reported. It is recommended that you use a stethoscope for checking bearing noises, because a bearing that sounds okay to the human ear may be producing chatter that will be revealed by the stethoscope.

Many contracts require that all equipment be installed and tested by factory-trained mechanics. If they are on your job, work with them to get a proper installation.

Once a machine has been test run and found to be okay, the bearings should run very warm, but not hot. You should be able to touch them without hurting your hand.

3.13. Bedrock

Bedrock is the solid rock underlying soil, shale, and other superficial rock. When the plans call for the foundation to rest on solid rock, they will require the removal of all loose material and that the surface be reasonably level. When the natural rock slopes, it is usually required that steps be cut to provide a level surface for the foundation. All structures should be founded on bedrock when possible, but much of the time it is too far down, thus piling, spread footings, or other methods are used.

3.14. Bell Holes

Bell and spigot pipe laid on the earth should have a bell hole dug as shown in Figure 3.11 so that the pipe will rest on the full length of the barrel rather than on the bell. After the joint is completed, the hole should be backfilled with compacted earth. Asbestos pipe with couplings should be treated the same way.

3.15. Bench Down

Bench down is a term used when an excavation is made in a series of steps (see Figure 3.12). The method is used in areas of unstable soil and in trenches where sewer and water lines are laid near each other. The water line will be on a higher level, so a leak in the sewer would not drain into it.

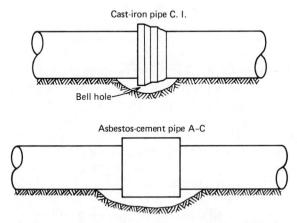

FIGURE 3.11. Bell holes should be dug for the bell or coupling of the pipe so that the body of the pipe will rest on the subgrade or compacted lining. If no bell hole is dug, the pipe will rest on the bell or coupling, and the pipe body will act as a beam when the trench is backfilled. The weight of the earth above the pipe may be enough to break it, especially the A–C pipe. After the body of the pipe has been full bedded, the bell hole should be backfilled with thoroughly compacted earth so that the support under the pipe is the same throughout its full length.

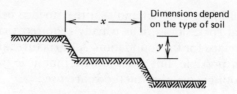

FIGURE 3.12. Bench-down excavation.

3.16. Bench Marks (BMs)

A bench mark is a point at which an elevation has been established for reference on a survey or during construction. The mark may be a stake, a spike in a tree, or a mark on a solid surface such as a concrete slab, or for permanent reference it may be a bronze disk. There will be the letters "BM" in the vicinity to help spot the place quickly. For temporary bench marks used during construction, the letters will be "TBM," for temporary bench mark. You must not disturb a bench mark, as replacement can be slow and costly.

3.17. Bentonite

Bentonite is a clayey substance mixed with water to make slurry (see Section 20.33).

3.18. Bituminous Concrete

Bituminous concrete (asphalt) is made by mixing graded sand and stone with hot (about 300°F) asphalt. The asphalt acts as a glue or binder, the same as Portland cement binds in Portland cement concrete.

The mixing is done at a central mixing plant and delivered to the job site in dump trucks. The trucks should have a heavy canvas sheet over the mix to retain the heat and exclude dust during transporting. The mix will be dumped into the hopper at the rear of the spreading machine (see Section 3.2). There will be a screw shaft in the bottom of the hopper to spread the mix evenly over the full width of the hopper. To do this, the screws should be covered with mix at all times when operating. When the mix gets low in the hopper, it may be necessary to use a hand shovel to spread the mix out over the screws.

The mix should be between 270 and 300°F when it is dumped into the hopper. If the mix is too cold, it will not spread evenly, and if it is too hot, the asphalt will run into small pools which will show up on the road surface and be sticky during hot weather and hard and slippery when it is cold. If there is an inspector on the job, he will reject any load that is not within the heat range. If cold mix gets onto the road surface, there will be uneven spots and even bare spots, and if it is too hot, shiny spots will appear. In either case, the area should be cleaned with a square-point shovel, and fresh mix should be placed before the area is rolled. Occasionally, the spots will appear after rolling starts, and if so, they should be treated as noted above.

There should be a pail of kerosene on hand into which the hand shovel should be dipped before and during use to keep the mix from sticking to it.

There will be adjustment screws on each side of the spreading machine to keep the mix at the proper depth as it is deposited on the roadway surface. The specifications will indicate the required compacted depth, and the loose mix must be spread at a depth that will give this thickness. The loose mix will be compacted by about one-third after it is fully rolled. If there is an inspector on the job, he will check the depth, but it is recommended that the foreman do his own checking. This can be done with a small metal rod, a 16-d spike, or an ice pick and should be done every few feet. If an inspector finds the final thickness to be less than required, he may order the whole area removed and replaced, and that would be expensive.

A few minutes after the mix has been spread on the roadway there will be one pass of the roller to make a "breakdown" pass, and then, after a few more minutes, the rolling will start and continue until the mix is fully compacted to the depth required. The exact timing will depend on the weather. The mix should be cooled to the point where the breakdown pass compresses it very slightly; it should not be so hot that the first pass makes a deep ridge. The rolling should start at the low edge and work towards the center of the roadway. Each pass should overlap the previous one by 50%. When the final rolling is complete, there should be no ridges in the surface.

The arrival of the trucks at the site should be controlled so that not more than two trucks are waiting to be unloaded. If more trucks arrive and the mix cools during the wait, the plant should be advised and either the trucks should be sent to another job or the loading should be held up. If there is a breakdown of equipment, the plant should be informed so the trucks can be held.

At the end of the day's run, the spreading machine must be thoroughly cleaned. All metal that comes in contact with the mix should have kerosene brushed or sprayed on and allowed to set a few minutes, after which the surface should be scraped clean by use of a square-point shovel and a wide-blade putty knife. After the surface has been thoroughly cleaned, it should be treated with a mixture of ¾ kerosene and ¼ light oil. The shovels should be cleaned and treated the same way (see Figures 3.13–3.17).

FIGURE 3.13. Asphalt paving. Plant-mixed bituminous concrete is dumped from the truck into the spreader (a Barber–Greene). A 10-ton, two-wheeled tandem roller compacts the material. The base course shown will be covered with 1 in. of top course containing finer aggregate.

FIGURE 3.14. Asphalt paving. A base, or binder, course 1½ in. thick is being laid over the old road surface. A top course ¾ in. thick will be laid over the base course. Notice the worker with a rake spreading the just-placed mix to make a smoother joint with the run previously laid.

FIGURE 3.15. Asphalt paving of bituminous concrete. One inch of a hot-mix bituminous concrete (trade name Armorcoat) is being placed over an existing pavement. Notice that the depth is thicker on the outer edge of the curve to provide some superelevation. A spreader for bituminous concrete is in the distance, and a two-wheeled tandem roller is in the foreground.

FIGURE 3.16. The truck is dumping bituminous concrete into the hopper of a spreader, which is placing an 8 in. layer of base course for the taxiway of an airport. The truck does not have a canvas cover over the body, as the mixing plant is only a short distance away. The mix has No. 2 and No. 3 crushed stone.

FIGURE 3.17. Hot asphalt being sprayed onto a roadway surface. The asphalt will be immediately covered with No. 1 or No. 1-A (small) crushed stone. The gallons per square yard is regulated by valves on the truck. The gallons of asphalt and the pounds of stone per square yard will be indicated in the specifications.

3.19. Blasting

Blasting is a dangerous trade and should not be undertaken by inexperienced personnel. Most states will require that the "powderman," who is in charge of the work, be licensed or certified. As a foreman, you will be concerned with the safety of the personnel and equipment under your control when blasting is undertaken near your area of work. There will be a warning of some kind before each blast, such as a shout of "fire in the hole," or the sounding of a horn, whistle, or siren, and on most jobs there will be a signal that all is clear to return to work.

When blasting a large area, such as in a quarry or on a hillside, the personnel and equipment will be moved out of danger. In smaller areas, the blast will be covered with blasting mats made of timber or woven rope or cable mats. If there is blasting in your work area, you must be alert for any falling debris. Should any worker or equipment be damaged, report it to your superior at once.

3.20. Blocking

Underground pipes such as water mains or force mains will require blocking or tying of some type to keep the joints tight at all bends or changes in direction.

THRUST BLOCKS

DRAWING NOT TO SCALE

TYPICAL DIMENSIONS* OF THRUST BLOCKS FOR HORIZONTAL BENDS, TEES & PLUGS OF WATER MAINS

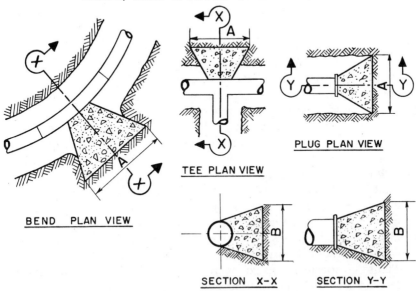

BEND PLAN VIEW

TEE PLAN VIEW

PLUG PLAN VIEW

SECTION X-X

SECTION Y-Y

PIPE SIZE	DEGREE OF BEND	DIMENSIONS A	DIMENSIONS B	PIPE SIZE	DEGREE OF BEND	DIMENSIONS A	DIMENSIONS B
12"	TEE OR PLUG	6'	3'	8"	TEE OR PLUG	4'	2'
12"	90	6'	4'	8"	90	5'	2½'
12"	45	5'	3'	8"	45	3'	2'
12"	22½	4'	2'	8"	22½	3'	1'
12"	11¼	3'	2'	8"	11¼	2'	1'
10"	TEE OR PLUG	4'	3'	6"	TEE OR PLUG	2½'	2'
10"	90	6'	3'	6"	90	3'	2'
10"	45	5'	2'	6"	45	3'	1½'
10"	22½	3'	2'	6"	22½	3'	1'
10"	11¼	3'	1'	6"	11¼	2'	1'

* *THRUST BLOCKS DESIGNED FOR 150 psi. HYDROSTATIC PRESSURE AND FOR BEARING AGAINST UNDISTURBED SOIL OF 1000 lbs/ft² BEARING CAPACITY.*

FIGURE 3.18. Thrust blocks. A sketch and tables similar to the ones shown are included in the specifications or on the plans of each project having pipe under pressure.

Thrust blocks are commonly used if the earth is solid enough to take the thrust. Measurements A and B will depend on the strength of the earth (see Figure 3.18).

3.21. Blueprints (Plans)

The plans of the work are usually referred to as blueprints or "plans." They are no longer "'blue," but are usually black lines on white paper. The plans show in detail what is to be constructed, and "plan reading" is a must for a good foreman. If you plan to become a foreman and do not fully understand plans, you should seek training. Many high schools now have adult-training evening classes where plan reading is taught; most community colleges do the same. If this method is not convenient for you, there are several home-study courses available, but you will have to discipline yourself to study regularly to get much from such courses.

Plans are a precision-drawn sketch or picture of what is to be constructed. There will be a "general" plan showing the overall structure, "sections" through the structure showing what it would look like if it were cut into several pieces, and "details," which are larger-scale drawings of the more complicated parts of the structure.

Plans and specifications go together, and when there is a contradiction between the two, the specifications govern (see Section 20.42).

On the more complicated structures, the engineer or architect will provide large-scale detailed sketches of the important parts as the work proceeds. Any such sketches must be studied and attached to your set of the plans with notes deleting the changes, if any, to the original. Many sketches make no changes, but rather give better details by being on a larger scale. You should make a note of when any such sketches are submitted to you.

There will be separate plans or drawings of the plumbing, electrical-heating, and air-conditioning systems if they are part of the project.

3.22. Boils (in Subgrade) (See Section 20.61)

When earth excavation proceeds to below the natural groundwater elevation, the upward movement of the water into the excavation can cause boils (see Figure 3.19).

Boils in the subgrade will cause an unstable condition which will extend down from a few inches to several feet, depending on the kind of material and the pressure of the water. Such conditions usually require the installation of well points or other dewatering systems (see Section 24.14). Sometimes the condition can be relieved by digging wells in the vicinity (see Section 24.15).

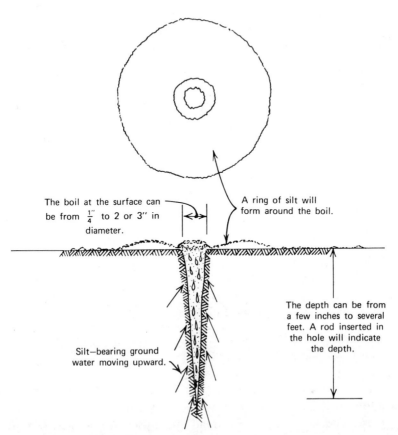

The boil at the surface can be from $\frac{1''}{4}$ to 2 or 3'' in diameter.

A ring of silt will form around the boil.

Silt—bearing ground water moving upward.

The depth can be from a few inches to several feet. A rod inserted in the hole will indicate the depth.

FIGURE 3.19. Boils in the subgrade. An unstable quicksand condition can be detected by patting the area with the foot to see if it has a jelly-like movement. However, the most frequent method is to observe boils in the subgrade. The upward pressure of the groundwater causes small springs, or boils, to develop on the surface, and quantities of silt and sand will be brought up and deposited in a ring around the hole. There may be one or several dozen boils, depending on the conditions at the time. If boils do develop, the subgrade can be disturbed for several feet down, and the area will have to be compacted after the boils are eliminated, usually by well points.

Sometimes the excavation will uncover a pocket of water under pressure which will cause boils for a few minutes until the pressure is relived. If it is within the excavated area there is no problem, but if it is in the subgrade, the unstable material must be replaced with a well-compacted proper material. If the pocket area is extensive and is in the subgrade, it can sometimes be stabilized by placing stones (2–6 in.) into the unstable material and forcing them down until the area becomes stable. This method is usually more economical, so if you encounter such a condition, it is recommended that you confer with your superior and the inspector.

If a very bad boiling condition is encountered, it is recommended that the area be backfilled immediately and left until the groundwater is lowered.

If you are excavating in an area of high groundwater, a careful watch should be kept to spot any boils as quickly as they are uncovered, and arrangements should be made to eliminate them.

Boils in earth with a high clay content may develop to a depth of several feet, while those in gravel and/or sand will usually be only a few inches deep. In either case, the unstable area must be consolidated before work proceeds.

3.23. Bolts (See Section 2.12)

There are many kinds of bolts on civil engineering projects.

Carriage bolts are used with wood. The head has a rectangular section that extends down into the wood to keep the bolt from turning.

High-strength bolts are used in place of rivets for the connections of structural-steel units. The specifications will usually call for the use of a "torque wrench" for tightening the nut (see Section 21.22). These bolts will stand a stress of over 100,000 psi.

Machine bolts are used to connect metal units and machine parts. They will have a square or octagonal head and nut and may require lock washers.

Rock bolts are used to hold loose sections of rock in place during excavation, especially in tunnels (see Figure 9.13).

Upset-end bolts have the thread end upset or enlarged so that the shank has the same diameter under the threads as for the rest of the bolt. This provides greater strength for the bolt (see Figure 3.20).

Construction Inspection: A Field Guide to Practice

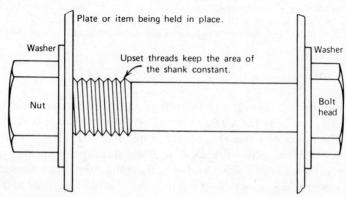

FIGURE 3.20. High-strength bolts. When high-strength bolts are used in tension, it is usually required that they have "upset" threads, which keep the shank area constant to develop the full strength of the bolt.

In areas of high humidity or moisture, the specifications will require that the bolts be galvanized or cadmium plated. These bolts must be handled carefully so as not to damage the coating.

Care must be taken to see that the nut on the bolt is drawn up as tight as possible without stripping the threads.

3.24. Bonds

Bonds for bidding on or completing a project area required on public works contracts and some private projects. They are furnished and approved before construction starts and are of no concern to the foreman.

3.25. Boring

Boring covers a wide range of procedures. Holes are bored in various materials to insert bolts, wood-frame structures are bored to install electrical and water lines, existing foundations are bored to insert anchor bolts, various materials are bored to obtain samples for testing, and borings are made through the earth to install conduits for water, sewer, and power lines. In areas where unstable earth overlays rock or other firm bearings, vertical borings are made, and the hole is filled with concrete to form piers.

Borings into structural frames must not remove so much material that the strength of the member is reduced below permissible limits. These limits will be indicated in the specifications or by the resident engineer. You should not, on your own, decide to bore into any structural frame. If it becomes necessary to make an unscheduled bore, confer with your superior before proceeding.

Power drills are used to bore holes, using highspeed for metals and low speed for other materials. Small holes can be bored in wood and other soft materials by use of a hand drill.

The cutting tool is called a "bit." The cutting point of the bit will slant at about 45 deg to the surface to be bored, except with the wood bit where the cutting end will have a knife-like edge that actually shaves off the wood. There will be a screw-like point that starts the hole and keeps the bit on line. When the bit size exceeds 1 in. in diameter, it is usually called an auger, particularly if used for wood or earth.

When boring into steel or other hard materials more than about ½ in. thick, there will be provisions to keep the bit cool by use of water, air, or other means. If the bit gets too hot, it will loose its "temper" or hardness and will no longer cut the material.

The bit for boring into concrete or masonry will have a case-hardened tip which is harder than the body of the bit. For drilling into rock, a "star

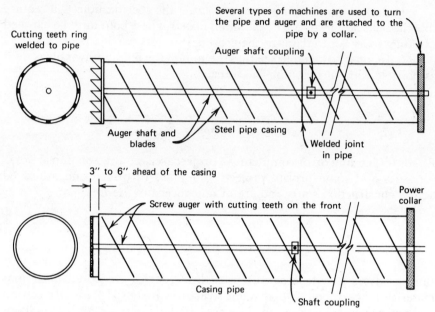

FIGURE 3.21. (top) Boring for casing. Several types of machines are used to turn the pipe and auger. They are attached to the pipe by a collar. The type that has a ring of teeth attached to the forward end of the casing cuts a hole slightly larger than the casing to reduce the friction on the casing as it is moved forward. (bottom) Another type of boring machine has the teeth attached to the auger, independently of the casing. With this system, the auger does the cutting and forces the loosened earth back. As the auger progresses, the casing is jacked forward to keep as close to the cutting ring as possible. This type cuts a hole slightly smaller than the casing pipe. As the casing is jacked forward, the forward edge shears off the earth, keeping the void outside the casing at a minimum. With either type, the auger can be operated while the casing is stationary to remove all the loosened earth from inside the casing.

drill" is used. The point will be shaped like a star with ridges that break up the rock like a chisel. For deep holes, the star drill will have a hole down the center of the rod through which air forces out the dust and rock particles. When boring, or drilling, several feet into rock to blast or to install rock bolts, the machine will rotate the rod, and for very deep holes, water instead of air will be used to cool the drill bit and remove dust particles.

When drilling into rock or other material to obtain a sample, the drill bit will be hollow like a pipe, and the cutting end will have case-hardened teeth or small diamonds embedded in it. These cut a circular trench, leaving a core in the pipe which, when removed, becomes the sample.

Boring through earth is done with a steel casing pipe, usually 18 in. or more in diameter. An earth auger inside the pipe casing will break down the earth and convey it back to the end of the casing for disposal. As the auger cuts into the earth and conveys it back, the casing is forced forward

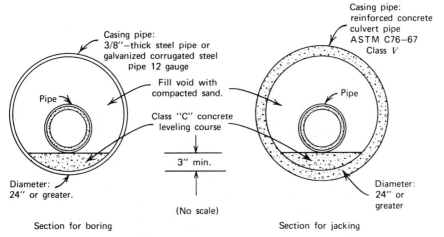

Section for boring   Section for jacking

FIGURE 3.22. Sewer pipe in casing for crossing under a highway or railroad. Boring jobs use steel casing pipe; jacking jobs usually use reinforced concrete pipe, although steel pipe can be used if the earth is soft enough. The void between the sewer pipe and the casing is filled with sand blown in under pressure; sometimes grout is used. The steel casing pipe eventually rusts out and leaves a large void if sand is not used. If concrete pipe is used, groundwater may seep in and float the sewer pipe.

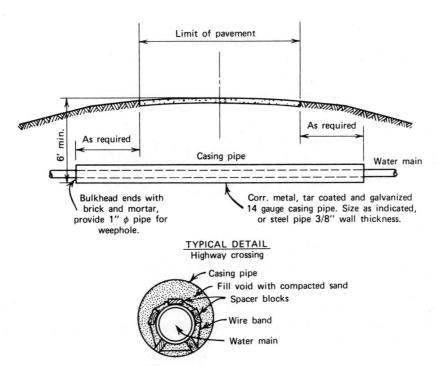

FIGURE 3.23. Highway-crossing detail. Notice the section showing wood blocks, which facilitate the sliding of the sewer pipe through the casing. Sometimes the two bottom blocks are greased.

FIGURE 3.24. Sewer pipe in casing under highway. The sewer pipe is in place within the casing. Notice the wooden bulkhead, which retains the sand fill. The reinforcing steel and forms in the foreground are for the foundation of the manhole.

FIGURE 3.25. A 36-in. steel casing pipe has been drilled under a roadway embankment. Sheeting and bracing will be installed, and lining will be placed on the subgrade before the sewer pipe is installed. A leveling course of concrete will be placed on the invert of the casing pipe to keep the sewer pipe on grade. After the sewer pipe has been installed, the rest of the casing pipe will be filled with sand and the ends will be bulkheaded.

so that the cutting device on the auger is never more than a few inches ahead of the casing. Some earth augers have an expanding head which cuts a hole in the earth which is just a little larger than the outside of the casing, thus reducing the friction and making it easier to force the casing forward. Another type of earth auger will have its teeth welded to the forward end of the casing, which rotates. The auger inside is used only to convey the loosened earth back for disposal. Here again, the teeth will cut an opening just a little larger than the diameter of the casing to reduce friction. Either method requires a special machine and a trained crew to operate it.

The biggest problem is keeping the casing on line and grade as it proceeds through the earth. To do this, the casing is set to line and grade on heavy timbers, which keep it in position as it starts to bore. The timbers will be long enough to accommodate 20–30 ft of casing; the longer the bed, the better the control. Bores of over 50 ft in length are difficult to control, especially with casing under 18 in. in diameter. If the boring proceeds forward at a slow, even pace in earth that is fairly uniform in texture, it will usually stay on line and grade for at least 50 ft. Beyond that, it is questionable. Obstacles such as tree roots, large stones, hard layers of earth, and buried pieces of timber can deflect the casing. If the casing pipe is 30 in. or more in diameter, a worker can crawl in and cut enough of the object to allow the casing to pass. The worker will usually use a small electric or air hammer for this work. If the casing is too small for the worker to enter

FIGURE 3.26. A type of boring operation. By this method, the steel casing, which has teeth welded to the forward end, is rotated and forced forward by the machine. A screw conveyor on the inside forces the loosened earth back to the rear for removal. (See Fig. 3.21.)

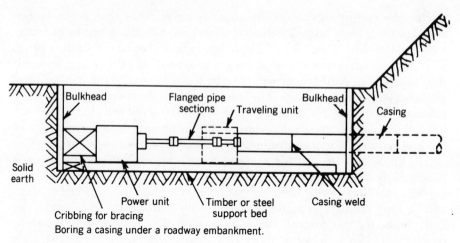

Boring a casing under a roadway embankment.

FIGURE 3.27. As the casing is moved forward, sections of the flanged pipe are removed and another section of casing is installed and welded to the preceding one. Some units will have the power attached to the casing and move forward by use of cogged rail on the support bed.

or the obstruction cannot be removed, it will be necessary to move to a new location.

For sewer or water lines, it is customary to install a casing pipe considerably larger than the sewer or waterline to allow room to adjust the line and grade (see Figure 3.21–3.26).

Vertical earth bores are made by a machine similar to that shown in Figure 3.27, only the auger is much larger, 18 or 24 in. or more. As the auger drills, the blades become loaded with earth, so the drilling is stopped, the auger is raised, and the rotation is reversed, which throws the earth off onto the surrounding ground.

3.26. Boring Log

A boring log is a record of the various types of soil encountered when making a test boring or test pit. It will show the design engineer the type of soil the footings will sit on so he can provide the proper size. The engineer or inspector will check to see that the soil actually encountered is the same as shown on the log. If not, he will confer with the design engineer to see if changes in design will be required. If you encounter soil conditions that differ from those shown on the plans, you should draw it to the attention of the inspector at once to avoid a possible delay (see Figures 3.28 and 3.29).

Depth	C.	N.	Sample No.	Sample Depth	Description of Material
		1/2	1	1.0'–	Brown moist loose fine medium SAND,
		2		2.5'	some silt
5.0					
		1/2	2	5.0'–	_____ 6.0'
		2		6.5'	Brown moist loose SILT and fine SAND,
					trace organic material
10.0					
		6/4	3	10.0'–	_____ 11.4'
		9		11.5'	
▼ WL					Brown moist medium dense coarse to
15.0					fine SAND and fine to coarse GRAVEL
		11/17	4	15.0'–	_____ 15.5'
		18		16.5'	Brown wet dense to medium dense
					coarse to fine GRAVEL and coarse to
20.0					fine SAND
		9/12	5	20.0'–	
		15		21.5'	
25.0					
		11/7	6	25.0'–	
		5		26.5'	
					_____ 29.0'
30.0					
		9/10	7	30.0'–	Brown wet medium dense fine SAND
		12		31.5'	and SILT
35.0					
		5/9	8	35.0'–	_____
		13		36.5'	
					Bottom of Boring 36.5'
40.0					

FIGURE 3.28. Test boring log. This is a reproduction of an actual log taken from the files of a project in Central New York State. The log shows where the samples were taken and gives a description of each sample. Notice the level of the groundwater indicated by WL.

3.27. Borrow (Borrow Pit)

Sometimes earth is brought to the site to fill in a low area, construct an embankment, or raise the elevation of the ground around a structure. Such earth is called borrow, and the place from which it is obtained is a borrow pit. The borrow pit may be several miles away or it may be on the job site of a large project.

Subsurface Investigation

Report No. L225-1-3-78

Client: _Highway Department_   Location of Boring: _See Plane_

Project: _Bridge Reconstruction_
New York   Date, start _1/4/78_ Finish _1/4/78_

Boring No. _B-1_ Sheet _1_ of _1_

Casing Hammer   Sampler Hammer

Wt. _____ lbs.   Wt. _140_ lbs.

Fall _____ in.   Fall _30_ in.

☐ _____ I.D. Casing
Ground Elev. _102.3_ ☒ $2\frac{3}{4}''$ I.D. H.S. Auger

Ground Water Observations

Date   Time   Depth   Casing at

No water noted in hole

Depth	Casing Blows/Ft.	Sample No.	Depth of Sample From	To	Type Sample	Blows on Sampler Per ____ Sampler O.D. ____	Depth of Change	Classification of Material F — Fine And — 35 – 50% M — Medium Some — 20 – 35% C — Coarse Little — 10 – 20% Trace — 0 – 10%	Depth
			0.0	0.2		Auger	0.2	Blacktop	
		1	5.0	6.5	SS	51-59-23		Brown cmf SAND; little mf GRAVEL; trace SILT (wet, non-plastic)	
		2	6.5	8.5	SS	12-7-8-11		Brown cmf SAND; little mf GRAVEL; little SILT (wet, slightly plastic)	
		3	8.5	10.0	SS	21-14-16		Brown cmf SAND; little mf GRAVEL; little ORGANIC Material (wet, slightly plastic	
		4	10.0	10.2	SS	13-100/0"	10.2	Ditto	
			10.2	10.7	AX	CORE		Bedrock 6" or 100% Recovery 3 pieces, 50% chips & fragments	
			10.7	12.7	AX	CORE		17" or 71% Recovery 10 pieces, 2 chips	
			12.7	15.3	AX	CORE		31.5" or 105% Recovery 16 pieces, 6% chips & fragments	
								Bottom Boring 15.3'	

SS — Split Spoon Sample.   Drillers _____ _____
U — Undis. Shelby Tube.
P — Piston Type Sample.

FIGURE 3.29. Subsurface investigation. This is a reproduction of another boring log of an actual project. Notice that no groundwater was present. Below the 10-ft level, rock was found, and a core drill was used for obtaining samples.

52

The specifications will indicate the kind of material that is acceptable. It may be required that the material at the pit be approved before hauling begins. If you are placed in charge of a borrow pit, you must keep close watch as the material is excavated to ensure that it will meet requirements. The character of the soil can change quickly, so it may be necessary to move to another area in the pit to obtain the quality and gradation required.

The trucks hauling the earth should be able to move quickly and easily into the pit, be loaded, and move quickly and easily out. This may require some roadwork to keep the trucks from bogging down. A small dozer is handy for this work. Trucks should be filled to capacity, but not overloaded to the point where earth will spill onto the highway. It may be required that you keep a record of how much each truck hauls, so each should have an easily read number.

When excavating at a borrow pit, the area should be left with a slope to provide good drainage so it will dry out quickly after a rain. Usually, this can be done by observing the surface, or stakes can be set along each side and the area can be sloped by use of grading sticks, as shown in Figure 8.4.

3.28. Box Beams

A box beam can be of reinforced concrete or structural steel. It will be shaped like a rectangle or trapezoid, with side, top, and bottom slabs. When made of concrete, the slabs will be from 4 in. to 1 ft thick, or if made of steel, it will have plates ½ in. or more in thickness. The center will be hollow and can be used to carry utility lines, sewers, or water mains. The sections will be placed by crane, and several sections may be placed side by side to make a wide roadway. The sections will be bolted or welded together. The top slab will usually have a roadway wearing surface placed on it (see Section 17.37).

3.29. Brick

Brick construction will be handled by the masonry foreman. Your crew may be assigned the job of bringing supplies to the masons and erecting and moving the scaffolding. In that case, you should follow the masonry foreman's instructions. Be sure that all scaffolding is well founded and secure from movement (see Section 20.6).

3.30. Bridges

Bridge construction is more difficult than most other types of projects because it is done over water, swamps, unstable areas, or other traveled ways. In many locations, the elevation of the water surface varies, and the speed of the flow causes problems. In winter, there are ice floes.

Bridges may be of reinforced concrete, structural steel, timber, or a combination of these. Long-span bridges, those over 300 ft, are usually steel. Secondary-road bridges are usually treated timber. Many spans of 100 ft or less are precast concrete.

The actual construction, such as driving piles, erecting falsework and forms, placing reinforcing steel, pouring concrete, and erecting structural steel, all varies little from other projects. However, when working over water or swamps, much of the equipment will be on barges, on a tramway erected along side the bridge, or on an overhead cableway, all of which will slow the progress. The foundations will be inside cofferdams or caissons where keeping the water out can be a problem. There is not the working room that is available on most other projects. Because of these problems, many contractors specialize in this work and have specially trained crews.

Some bridge foundations require the use of pneumatic caissons, which is another specialized field (see Section 4.1).

In many locations where the water is shallow or swampy along the shoreline, the bridge will be built only over the navigable portions of the river or where the water is deeper, and the roadway embankment will connect the bridge with the shore. If the flow in the river is swift, particularly at high water, riprap will be installed on the slopes of the embankment (see Section 19.13).

In some areas where the foundation rock is seamy, it may be impossible to dewater the caisson or cofferdam, in which case the lower layers of

FIGURE 3.30. A railroad bridge built of creosoted timber. Each bent has six driven piles with four vertical and the two outside battered. The pile cap is 12 × 12 in. and the stringers, or beams, are 6 × 12 in., placed 6 in. apart for the full width of the deck. There is a 6 × 8 in. curb on each side, set on 3 × 6 in. blocks set about 18 in. on center. Note that the dumped riprap at the near end has not been carried around the side as it should have been. When such bridges are built of sound, pressure-creosoted timber, they will be maintenance free for many years.

FIGURE 3.31. Bridge shown was built under the supervision of the author. Figure 7.1 shows part of the falsework for this bridge.

concrete may be placed by tremie (see Section 21.27) or by pressure grouting around the perimeter (see Section 8.19).

For reinforced-concrete bridges, the mixer will usually be on the shore, with the concrete being delivered to the point of deposit by buggy over walkway, by bucket over a narrow-gage railway on a trestle, by overhead cableway, or by pumping through an extended pipeline. On large projects, the mixer with the aggregate may be on barges, with the concrete being placed by crane and bucket or by pumping.

Small structural-steel bridges will have the beams and girders placed by crane from the shore, and longer spans will have one or more cranes on barges. Steel trusses will be erected on falsework, be cantilevered from the supports, or be erected by floating cranes.

The labor foreman and his crew will work on preparing the job site, constructing and removing cofferdams and falsework, and placing reinforcing bars and concrete. They will assist in setting and removing concrete forms with the carpenters, in handling the barges and in moving structural steel units from the storage yard to the point of erection, and they will assist any of the trade groups in handling their material.

A bridge or viaduct over another road or a railway will require special attention to traffic below. The traffic itself must be interrupted as little as possible, and care taken not to drop anything. Some projects will require a net under the work to catch any falling object and to protect a falling worker (see Figures 3.30 and 3.31).

3.31. Brooming

Brooming can mean two things. The surface of fresh concrete may be roughened by drawing a broom across it. The brooming may be done to provide a better bond for concrete placed on top, or it may provide a nonslip surface on a floor or walk.

Brooming is also done to clean the floors of a structure. The specifications require the rooms to be "broom clean."

3.32. Budgets

Occasionally, a project is constructed on a strict budget, in which case it will be determined how much funding is available to cover each section of the work. You will be advised not in dollars but in time available for labor and equipment to do the work assigned to you. You must keep accurate and complete records of each item so that at the end of a period the dollars spent can be compared to the progress made and adjustments can be made where necessary.

3.33. Builder's Hardware

Builder's hardware includes door hinges (butts), knobs, locks, closers, bumpers, night latches, etc. It also includes window locks, balances, bathroom fixtures, doorbells, house numbers, light fixtures, and almost anything else made of metal and used to construct a house. These items are purchased through hardware stores or builder-supply houses.

The plans and specifications will indicate the kind, quality and location of the various items. The foreman's job is to see that each is properly installed and protected until the building is turned over to the owner.

3.34. Buildings

The foreman on a civil engineering project will be concerned with powerhouses, pumping stations, treatment plants, and control buildings. Most other buildings will be constructed by "building contractors," those who specialize in buildings.

The foreman will be involved in layout, excavation, foundations, backfill, floors, walls, partitions, roofs and roofing, electric, heating, air conditioning, plumbing, painting, insulation windows, doors, and so on. All of these are discussed under separate headings (see Figures 3.32 and 3.33).

FIGURE 3.32. The front of the control building for a large sewage treatment plant. The basement contains the bar screen and grit chamber. On the first floor are the office, the sludge driers, lime-solution mixers, and a garage. The top floor has a laboratory, lime containers for the mixers below, and a storage room. Note the loudspeakers on the post. These are for sounding an alarm when some unit in the plant malfunctions. The alarm is activated from the control panel.

FIGURE 3.33. The front of a group of four circular tanks with a control building between them. Three of the tanks are digesters and one is a gas holder. The digesters have domed concrete roofs, and the gas holder has a floating steel cover. An underground pipe gallery runs from the main control building to the digester tank control room.

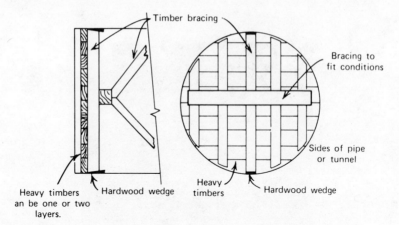

Timber bracing

Bracing to fit conditions

Sides of pipe or tunnel

Heavy timbers

Hardwood wedge

Heavy timbers

Hardwood wedge

Heavy timbers an be one or two layers.

FIGURE 3.34. Timber bulkhead. Bulkheads are temporary plugs used to close openings such as the end of a pipe or tunnel. The size and strength of the bulkhead depend on the situation. If the bulkhead is used only to keep people and animals out, it does not have to be heavy. However, if, as in a tunnel, the bulkhead must resist pressures from earth, it must be strong enough to do the job. Critical bulkheads are designed by an engineer, and the details are shown on the plans.

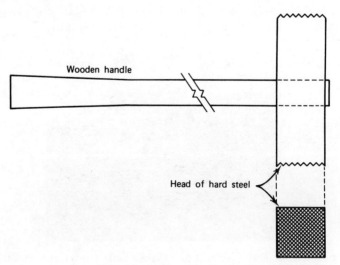

Wooden handle

Head of hard steel

FIGURE 3.35. Bushhammer.

58

3.35. Bulkheads

On civil engineering projects, bulkheads are temporary walls set up to retain earth during construction. In tunnels, a bulkhead is placed at the working end of the tunnel when work has to stop for a period of time (see Figure 3.34). The bulkhead should be constructed so that it can be quickly installed and removed, yet is strong enough to withstand the earth pressure that might come against it. The bulkhead is constructed of heavy timber with sufficient braces to hold it in place.

A casing pipe through which a sewer or water line is installed will have a brick bulkhead installed at each end, and if a tunnel or mine shaft is to be abandoned for a length of time, brick or masonry bulkheads will be installed in them (see Figure 3.34).

3.36. Bulking

Dry sand and most grains (e.g., wheat, corn, or rye) will bulk when moisture is added. That is, the volume will increase so that 1 ft^3 of dry sand will become about 1.15 ft^3.

3.37. Bushhammers

The bushhammer (see Figure 3.35) is used to break up the surface of concrete for architectural effects. The surface will be penetrated about $\frac{1}{16}$ in., exposing the aggregate. If properly done, it produces a pleasing effect. Care must be taken to see that the same pressure is applied for each blow of the hammer, and the surface of the hammer must be kept clean. The end panels of bridge handrails and the cornerstones of buildings are often bushhammered (see Section 20.2).

4

4.1. Caissons

There are two kinds of caisson used on construction: open-air and pneumatic.

Open-air caissons are constructed of concrete, timber, or metal boxes, open at the top and bottom, in which excavation is carried on until reaching solid support which will carry a bridge pier, an abutment, or the foundation of a structure (see Fig. 4.1). They may be in a river, lake or water-bearing soil. As the excavation proceeds, the caisson is lowered until the desired elevation or support is reached. Sometimes the support is bedrock or other material that will support the loads. Sometimes the caisson is lowered to a predetermined elevation and piling is driven the rest of the way. After the foundation has been constructed, the caisson may or may not be removed. If in a navigable stream, the caisson will be removed, at least in situations where it will interfere with navigation.

With timber construction, the individual planks may be driven down by jackhammer as the excavation proceeds, or a timber box may be constructed ahead of time and placed in position before excavation starts. Such boxes are heavily constructed, and weights of concrete blocks, sand, or material are used to help force the box down as the excavation proceeds. If the work is in an area of high ground water, the excavation is sometimes carried down to just above groundwater before the caisson box is placed.

If the area has high groundwater, the cracks between the planks will have to be caulked with oakum or jute. (See Section 4.13.)

Pneumatic caissons are open only on the bottom and are under compressed air to keep the water out (see Figures 4.2 and 4.3). These caissons can be of timber, concrete, steel, or a combination. They are usually fabricated

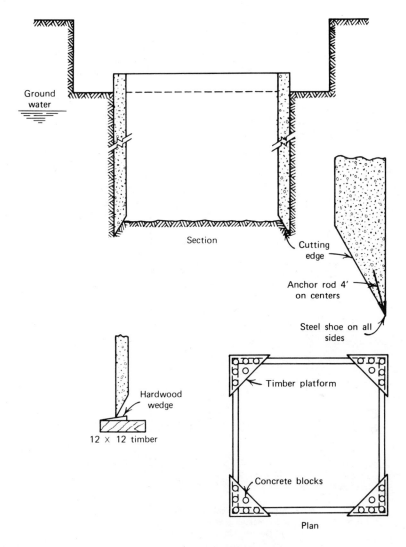

Ground
water

Section

Cutting
edge

Anchor rod 4'
on centers

Steel shoe on all
sides

Hardwood
wedge

12 × 12 timber

Timber platform

Concrete blocks

Plan

Excavation in open air caisson

FIGURE 4.1. Open-air caisson. Open-air caissons are usually built of concrete if they are rectangular, although they can be made of heavy timber. The circular ones can be made of heavy steel pipe or concrete pipe of large diameter.

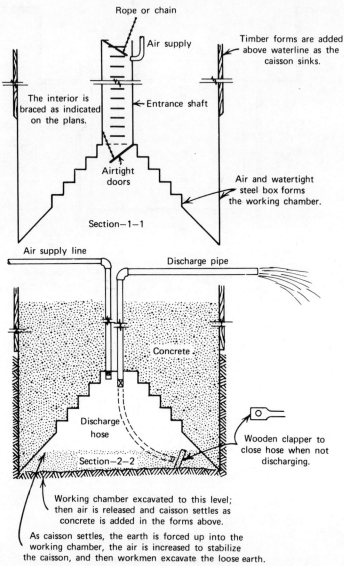

Rope or chain

Air supply

Timber forms are added above waterline as the caisson sinks.

The interior is braced as indicated on the plans.

←Entrance shaft

Airtight doors

Air and watertight steel box forms the working chamber.

Section—1—1

Air supply line

Discharge pipe

Concrete

Discharge hose

Wooden clapper to close hose when not discharging.

Section—2—2

Working chamber excavated to this level; then air is released and caisson settles as concrete is added in the forms above.

As caisson settles, the earth is forced up into the working chamber, the air is increased to stabilize the caisson, and then workmen excavate the loose earth.

FIGURE 4.2. Pneumatic caisson. The pneumatic caisson is a closed box tight enough to retain air under pressure. The illustration shows a caisson in cross section. This box was 22 ft wide and 66 ft long. It had a working chamber 9 ft high at the center. The workers went through approximately 18 ft of water and 25 ft of earth to reach the bedrock. To use a pneumatic caisson, the working chamber is excavated to about the level of the bottom of the caisson, and the air pressure is reduced. The caisson settles, forcing the earth up into the chamber. The air is again increased in pressure as required, and the workers (sand hogs) remove the loose earth by blowing it out through the discharge hose. Rock and large stones are removed by a bucket passed through the entrance chamber or on a large caisson through a separate air lock. As the caisson settles, timber forms are added on the top to hold the concrete, which is added to keep the pouring just above the water level.

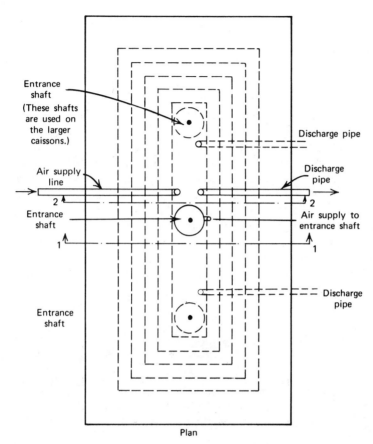

FIGURE 4.3. Pneumatic caisson. This illustration shows a plan view above the steel box. The center entrance shaft is used by workers. The two end shafts are used for removal of rock and stone and to fill the working chamber with concrete when the excavation is completed.

on shore or on a barge and are launched when completed, towed into position, and allowed to settle into place. The photos show a 22 × 66 ft steel working chamber with timber walls above, which will be filled with concrete as the unit settles.

Most pneumatic caissons are carried down to bedrock or hardpan. For bridge piers, a key is cut into the rock to increase the friction against sliding.

The workers in pneumatic caissons are called "sandhogs" because they kneel in the sand, mud, or other loose material astride a heavy hose and with a wooden paddle pull the loose material to the end of the hose where the air pressure forces it out of the caisson. The hose will be 4–6 in. in diameter and reinforced with steel ribs to prevent collapse from the air pressure. Excavation too heavy or too large to be blown out through the hose is hoisted out by buckets operating through an air lock. These air locks operate fast and with quick changes of air pressure, so no workers are inside them.

As the excavation proceeds and the working chamber is lowered, forms are added to the top, and the manlocks, bucket locks, and air supply and discharge lines are all extended upward to keep them above the concrete.

The workers enter the caisson through a man lock, which is a heavy steel pipe with airtight doors at each end. There will be steel steps similar to manhole steps welded to the side of the pipe. As the workers enter, the top door will be open, while the bottom one will be held shut by the pressure within the working chamber. When all are in the lock, the top door will be shut and sealed tight. The air pressure in the lock will be slowly raised to that in the working chamber, and when the pressure is equal the door will open. To leave the working chamber, the operation is reversed. The top door will be sealed shut and the bottom door open so the pressure in the pipe will be the same as in the working chamber. When the workers are in the pipe, the bottom door will be closed and sealed. The pressure in the pipe is slowly lowered until it equals the outside atmospheric pressure, at which time the top door will open and the workers will climb out. The

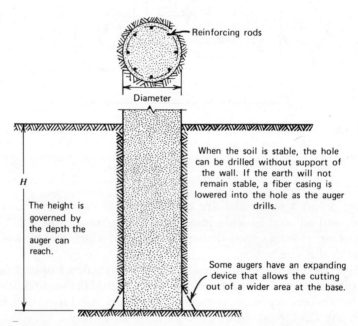

Reinforcing rods

Diameter

H

The height is governed by the depth the auger can reach.

When the soil is stable, the hole can be drilled without support of the wall. If the earth will not remain stable, a fiber casing is lowered into the hole as the auger drills.

Some augers have an expanding device that allows the cutting out of a wider area at the base.

FIGURE 4.4. Drilled caisson. When the soil is stable, the hole can be drilled without supporting the wall. If the earth is not stable, a fiber casing is lowered into the hole as the auger drills. Some augers have an expanding device to allow cutting out a wider area at the base. On holes of large diameter, a worker can be lowered into the hole to cut out the enlarged base by hand. The wider area allows for greater bearing area. Drilled caissons are used to provide footings for columns or grade beams when excavation to stable soil is not required for a basement.

FIGURE 4.5. Half of a 60-ft-long working chamber of a pneumatic caisson. The other half will be fabricated, and the two will be welded together, lowered into the water, and floated to the pier site. (See Figure 4.2)

regulation of the air pressure in the pipe and the time the workers must spend in the pipe during pressure changes are regulated by law and depend on the pressure being maintained in the working chamber. The air pressure depends on the composition of the soil around the caisson. Impervious soils like clay hold the air at considerable depth, while sandy and gravelly soils release the air at much shallower depths. (See Figure 4.4.)

Care must be taken to keep the caisson vertical, as any tilting will allow the soil on the high side to loosen so that air can escape, reducing the pressure in the working chamber. Correcting a tilt is a difficult job and is best learned by observing such an operation.

"Working under air," as the work is called, is a specialized field and a young man's job. Very few workers over 35 or 40 years of age undertake such work, as it is hard on the body. It is a high-paying job, and if you aspire to become a foreman, it is recommended that you work at least a year under a foreman of wide experience.

The caisson shown is for a large bridge pier (see Figure 4.5). They can be smaller, with some being made of large-diameter reinforced-concrete pipe or heavy-gauge steel pipe. Sections of pipe are added as the work progresses downward.

4.2. Calcium Chloride

Calcium chloride is used to control dust and to remove snow and ice. It is very corrosive, so care must be used to keep it from contact with metal. It should be spread as lightly as possible and still do the job. Any excess on the road will be picked up by the truck tires and thrown against the metal body.

Some contractors ask to use calcium chloride in the concrete to avoid freezing, but most specifications will forbid the practice. Some specifications allow other chemicals.

4.3. Cambers

The upward arching or curving of a beam or girder is called a camber. When a beam or girder is loaded, it will give so that if it is straight before loading it will bow down after loading. The camber will keep the beam straight or slightly above after loading. This will add to the strength of the beam as well as look better.

4.4. Canals

Canals range from small irrigation canals in the Southwest to large ship canals such as the Welland Canal between Lakes Erie and Ontario. Modern canals are lined with Portland cement concrete, bituminous concrete, or rubberized fabric. In former times, the lining was clay in 4–6 in. layers.

FIGURE 4.6. An irrigation canal with concrete lining. Note the two methods of construction. One is a precast slab set in place and the joints sealed; the other is poured-in-place concrete with wire mesh reinforcing. The bottom of the canal is a continuous poured-in-place, mesh-reinforced concrete slab.

FIGURE 4.7. The same canal as in Fig. 4.6, with a highway bridge crossing.

As foreman, you could be involved in excavating, shaping, and lining the canal and in installing control devices, which can run from simple piping and valves to overflow weirs, spillways, or locks for boat traffic.

For lining, see Section 13.18. The other items are covered under separate headings elsewhere in the book. (See Figures 4.6 and 4.7.)

4.5. Cant Hooks

A cant hook is a wooden pole about 4 ft in length and 2–3 in. in diameter, with a steel spike in the lower end and a steel hook on one side. The unit is used to roll logs or heavy timbers into place.

4.6. Cantilevers

A cantilever is a beam or other part of a structure that extends beyond its support. (See Figure 4.8.) A balcony, bracket, or overhang of a roof are all cantilevers. Some bridge spans cantilever from the piers and join at the middle of the span. For structural steel, the bolts must be up tight and should have lockwashers.

Usually, a concrete cantilever slab is not poured with the wall. A recess is left as the wall is poured, and the cantilever reinforcing bars are cast with the wall. Figure 4.8 shows how the bars are placed.

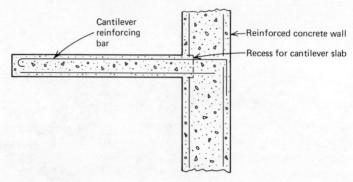

FIGURE 4.8. Usually the cantilever slab is not poured with the wall. As shown, a recess is left in the wall, with the slab reinforcing bars cast with the wall concrete. Note the cantilever reinforcing bar is in the top of the slab, and this must be carefully checked before the wall concrete is poured. If the bar is lower than specified, the strength of the slab will be drastically reduced.

4.7. Carpenters

Carpenters on civil engineering projects erect batter boards, falsework, scaffolding, forms for concrete, wooden framework, and other wood construction on the project. They will work under their own carpentry foreman, or if there are only a few, they may work under a general foreman.

4.8. Cast-in-Place Concrete

Cast-in-place concrete is poured into forms, set so that when the forms are removed the concrete will be in its proper position in the structure. (See Section 17.37.)

4.9. Cast-Iron Pipe

Cast-iron pipe is used for all sanitary lines within a building, except the smaller ones may be made of copper. If copper is used with cast iron, a plastic adaptor must be inserted between the two. (See Section 4.12.) The cast-iron joints will usually be bell and spigot, with the joint being sealed with jute and hot lead. The jute will be rammed tight into the back of the joint, and the remainder will be filled with hot lead. After the lead has cooled, it will be caulked by use of a caulking iron and a hammer to seal it tight.

Cast iron is also used for water lines and sometimes for sewers. The joints may be bell and spigot, as noted above, or mechanical, with flanges and bolts. The mechanical joints will have rubber or composition gaskets. The bell and spigot may have gaskets or jute and lead. The valves and hydrants on water lines will be of cast iron and may have bell and spigot or flange and bolts.

The cast iron may be regular or extra-heavy (EH) strength. The specifications will indicate where each strength is to be used.

Sometimes the pipe will have cracks or flaws that are hard to detect by observation, but which will leak when the line is under pressure. It is recommended that all pipe lengths be "rung" before being placed in the ground. A pipe is "rung" by suspending it with a rope or cable sling and striking it with a hammer. If the pipe if sound, you will hear it ring like a bell. If there is a crack or flaw, you will hear a dull thud. It is much more economical to spend the time to check the pipe before installing it than to dig it up later to repair a leak.

4.10. Cats

When heavy earth-moving equipment first came on the market, most of the caterpillar-type bulldozers were manufactured by the Caterpillar Company of Peoria, Illinois, hence the name "Cat." Now the name Cat is used quite often when referring to any dozer. There are several makes on the market. The Caterpillar Company designates its dozers as D-4, D-6, D-8, and D-10, depending on the size.

4.11. Catch Basins

Catch basins are placed along the curbline of a paved roadway or in the ditch in areas with no curbs. The basins are constructed of concrete or masonry blocks. They will have a cast-iron grate for the storm water to enter the basin (see Fig. 4.9). The basin will have a stilling area below the outfall pipe to allow the grit and other solids to settle before the storm water flows into outfall pipes to be carried to the point of disposal. A baffle or trap will keep the suspended solids from entering the outfall.

4.12. Cathodic Protection

Cathodic protection takes advantage of the fact that when two dissimilar metals are connected one will corrode and the other will not. An anode metal will corrode, a cathode metal will not. The system is used to protect steel pipe when installed underground by connecting pieces of anode metal at intervals along the line. A long line may have a DC current supplied by batteries. For water tanks, the system is suspended in the water. The plans will outline the system required.

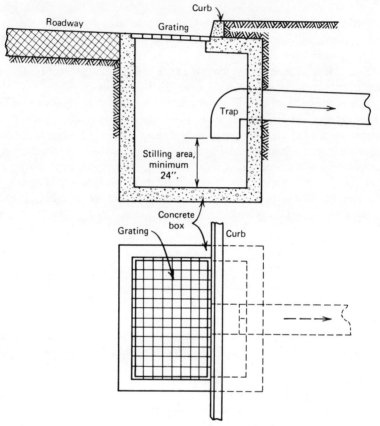

FIGURE 4.9. One type of catch basin. The catch basin is placed at the curb or in the centerline of a ditch. The stilling area allows the mud and silt carried by the storm water to settle and not get into the sewer. Maintenance crews remove the silt after a storm.

4.13. Caulking

The caulking, or sealing, of the cracks between the planks of sheeting or in the hull of a boat is done by braiding strands of oakum into a rope and driving it into the cracks with a caulking tool. This tool is somewhat like a chisel, only the point is not sharp. The oakum may be treated with bituminous material, which makes it more water resistant and makes it stay in place better. The caulking tool can be driven with almost any kind of hammer, although a wooden mallet is customary. Such caulking, when driven in tightly, will resist considerable water pressure.

A caulking compound similar to putty is used to seal the cracks in window and door frames and the walls of a building. This compound is a little softer than putty and is applied with a caulking gun. This gun applies a narrow, smooth bead over the crack and will seal out rainwater and air. The bead around doors and windows should be of even width for appearances.

4.14. Cement (Portland)

Portland cement is used to make concrete. The cement is the glue that binds the sand and stone (aggregate) together. The basic mix is 1–2–4, that is, one part cement, two parts sand, and four parts stone (or gravel). The exact mix for the various parts of a project are indicated in the specifications. Portland cement comes in five types. Type I is for general purpose use. Type II is a modified cement for use where the heat of hydration may be a problem and where additional protection for sulfates is desired. Type III is high-early-strength cement for use where full-strength concrete is required in a short time. This cement will produce a concrete as strong in 3 days as Type I will produce in the usual 28 days. Type IV is a low-heat cement used for mass concrete, such as in dams or large foundations. Type V is a sulfate-resistant cement. The specifications will indicate where the various types are to be used.

Today most concrete is delivered to the job site by "Redi-mix" trucks. (See Section 4.35.) Large projects may have their own on-site mixing plants, in which case the cement is delivered to the site either in bags or in bulk. In either case, it must be kept dry until used.

The strength of the concrete depends on the ratio of water to cement at the time of mixing, which varies from $3\frac{1}{2}$–5 gal/sack (1 ft^3) of cement. The exact water–cement ratio (W/C) will be specified.

4.15. Cement-Lined Pipe

A cement lining may be required for cast-iron pipelines when they are to carry certain acids or chemicals. It will also be used when a cast-iron pipeline is to carry water containing chemicals that would leave deposits on the unprotected pipe. The cement lining will be about 1/16 in. thick. Careless handling of the pipe can damage the lining, causing a loss of protection for the pipe.

4.16. Cesspools

When a regular sanitary sewer system is not available, one way of disposing of sanitary waste is shown in Figure 4.10. A study of the figure will give you an idea of the construction. It is recommended that guard posts be installed on all sides to keep heavy equipment off.

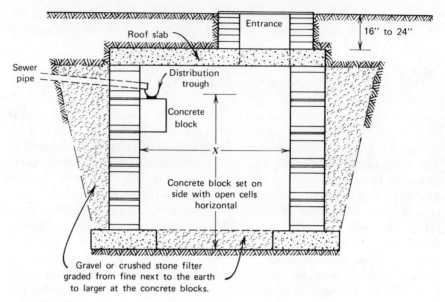

Gravel or crushed stone filter
graded from fine next to the earth
to larger at the concrete blocks.

FIGURE 4.10. One type of cesspool or leaching basin. A cesspool may be round, square, or rectangular. The distance X depends on the area of filter required and is shown on the plans. The distribution trough is set level and runs a length equal to half the inside perimeter of the basin. The ends of the trough are plugged. An A–C pipe sawed in two is recommended for the trough. Gravel or crushed stone is placed outside the basin to filter the water and keep the earth from slumping into the basin. The sewer pipe carries effluent from a septic tank or treatment plant.

4.17. Chaining Stations

The control stakes set parallel with the centerline of a pipeline, roadway, or other job that runs for some distance are referred to as chaining stations. The name is a holdover from the days when the measurements were made with a surveyor's chain. The chain, being of wire links, was 66 ft long. Today the stakes are set at 100 ft by a surveyor's steel tape. The tape may indicate feet and inches or feet and tenths of a foot. An inch equals 0.08333 ft. When recording your work in reference to the stakes, you must check to see if the measurements are in inches or tenths of a foot.

4.18. Chairs

The metal devices that support the reinforcing bars when concrete is being poured are called chairs. The chairs may be individual for large bars, a few inches long for supporting bars in beams and girders, or several feet long for use in concrete slabs. The chairs will be constructed to hold the bars at

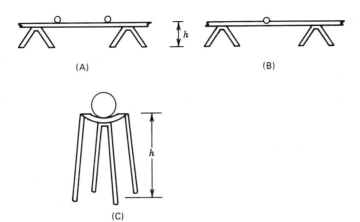

FIGURE 4.11. Chairs (reinforcing bar supports). Type-A chairs come in long lengths, and the reinforcing bars can be wired to the rod at any distance center to center. Type-B chairs come in long lengths with a notch for holding the reinforcing bar placed at specified distances. Type-C chairs are individual chairs for use with large-diameter reinforcing bars. The height h of all types will be specified to hold the bars at the required distance from the form or subgrade.

the proper height in the concrete and at the proper distance apart in beams and slabs. The bars will be wired to the chairs to hold them in place while the concrete is poured. (See Figure 4.11.)

4.19. Chalk Lines

Chalk lines are used to mark the line for building partitions, setting equipment, aligning bearing plates, or anywhere a temporary line is needed to mark a location. The line is similar to a mason's line. It is on a spool contained within a metal or plastic container which is filled with chalk, usually blue, that adheres to the line as it is unreeled. The line will be pulled taut between two points on a straight line on the floor. The line will be raised slightly and allowed to snap back, causing the chalk to mark the line on the floor.

4.20. Change Orders (See Section 6.22)

Change orders are issued when there is to be a change from what was originally specified and shown on the plans. The change can be for materials or a rearrangement of the same materials, or both. Proper change orders will include instructions on what is to be added, deducted, or rearranged and the increase or decrease in costs. As foreman, you should not make changes without a written order. If you are given a verbal order to make

CHANGE ORDER NO.: #   DATE: APRIL 10, 1979

CONTRACT NO.: 6

PROJECT: _____ SANITARY DISTRICT.   FILE: 55.27.

OWNER: _____ Department of Sanitation.

CONTRACTOR: _____

AUTHORIZATION IS HEREBY GRANTED FOR THE FOLLOWING CHANGE ORDER:

Description of Change Order:

        Substitute Cast Iron Pipe for Asbestos-Cement Pipe between MH#35 at Sta.
        19^+15 and MH#37 at Sta. 26^+38 on Smith Rd. Involving 715 linear feet of Pipe.

Reason for need for Change Order:

        New Industrial Park under construction will change traffic pattern to involve many
        large transport trucks daily.

Change Order Cost: _ _ _ _ _ _ _ _ (Per attached Itemized Breakdown) _ _ _ _ _ _ _ _ _ _ _ _ $3,585.00

Revised Contract Cost: _$5,606,736.51

        All work to be down in accordance with the applicable provisions of the
Contract and Specifications.

APPROVALS:

OWNER: _____   DATE: _____

CONTRACTOR: _____   DATE: _____

ENGINEER: _____   DATE: _____

      NOTE: Contractor would submit an itemized breakdown of the costs involved.

FIGURE 4.12. Sample change order.

a simple change, you should note such an order on your daily report, stating who gave the order and the time of day. Also, you must be sure the person giving the order has the authority to do so (see Fig. 4.12).

4.21. Checklists

When a project or a section of it nears completion, the engineer or his inspector will compile a checklist (punch list) of items of the work that must be repaired, adjusted, or corrected before the job will be accepted. You, as foreman, may be given a copy of the list with instructions to correct all or a certain part of the list. The list may be long or short, depending on the quality of the work done the first time. All items on the list must be corrected and the work must be reinspected before a certificate of completion is issued.

    Some contractors will do a minimum of work on the easy items on the list and call for a reinspection, hoping to get by with the least effort, but most engineers will not let them get away with it. You will have to be guided by the instructions of your superior.

A checklist could include, but not be limited to, the following items:

1. The ease with which doors and windows open and close
2. The fit of doors and windows
3. Whether switches control the lights and equipment indicated on the plan
4. Whether control panels operate the equipment as indicated on the panel
5. Whether equipment noise is within allowable limits
6. Operation level of float switches
7. Correct direction of flow indicators on pipes
8. Correct color coding on process pipe
9. Correct operation of ventilation and air conditioning
10. Required laboratory equipment accounted for and properly stored
11. Required office supplies on hand and properly stored
12. Operation of valves in proper direction
13. Correct operation and location of all alarms
14. Screens and storm windows supplied as required
15. All culverts and drainage ditches open and operating as required
16. References to underground facilities in place as required
17. All keys, including the master key, accounted for
18. Embankment trimmed to the proper lines and all eroded areas corrected
19. Topsoil and seed applied as required
20. All walls (both inside and out) and all floors clean
21. All trash removed from the site

4.22. Cherry Pickers

Cherry pickers are cranes with solid, extendable booms. They are smaller than the standard crane of the same capacity and are very maneuverable. As the boom retracts, the unit can move in congested areas. It will have outriggers that extend out at each side to stabilize the unit and that retract for moving. (See Figure 4.13.)

4.23. Chimneys (Smoke Stacks)

A chimney may be of masonry with flue lining, steel, or a combination of these two. Quite often the lower half will be of masonry with the top half of steel. If the top is steel, it will be attached to the masonry by heavy anchor

FIGURE 4.13. Cherry picker. A cherry picker is a small, solid-boomed, easily maneuvered crane. The boom will retract and sit level over the machine. When in use, the boom can be raised to about a 75 deg angle and can be extended to about three times the length of the lower section. The larger cherry pickers have outriggers that extend out about 5 ft to help stabilize the unit. They retract when the unit is moved.

bolts and may require guy wires at the top to stabilize the stack in a high wind. Chimney construction requires trained crews, but construction foremen may be assigned to handle supplies, erect scaffolding, and otherwise assist the erecting crews.

The anchor bolts must be tightened evenly, and the guy wires will have turnbuckles to equalize the tension on all lines. There will be an angle at the bottom of the steel section with holes for the anchor bolts. These holes should be about twice the diameter of the bolts to allow for temperature changes.

4.24. Chinking

Chinking is the process of placing small pieces of broken stone between larger stones used for riprap. (See Figure 19.9.) The small pieces can be spread over the joints and driven into the joint with a heavy sledgehammer. They should be driven in tight to hold the larger stones in place and to provide a tight surface to resist erosion during high water periods. The broken stone pieces are better than gravel, as the sharp edges will lock the mass together.

Chinking is also used when constructing a thick roadway base where No. 2 and No. 3 stone, or larger, is spread in 8–10-in. layers and rolled.

After rolling, the small broken stone is placed, swept into the voids between the larger stone, and rolled. The rolling will break up the pieces and force them down into the mass.

4.25. Chlorine

Both water purification and sewage treatment plants use chlorine in their processes. The chlorine may come in liquid or gas form and will be contained in metal cylinders or tanks, depending on the size of the operation. The chlorine will usually be delivered only after the project has been completed. If containers are stored on the site and should develop a leak, advise your superiors, but do not go near the area without a proper gas mask. If any of your crew should inhale escaping chlorine gas, send them immediately to a capable first-aid or emergency medical center.

4.26. Claims

Unanticipated problems on a construction project may result in extra work for the contractor, for which he will be entitled to extra compensation. Many such claims are rejected because they are not well documented. If your crew runs into a problem which you believe could cause extra work, you should confer with your superior. If it is determined that there could be a basis for a claim, you should keep accurate records of what was done, where, and why, listing labor, equipment, and materials used.

4.27. Clay Pipe

Clay pipe may be glazed or unglazed; however, today most unglazed clay pipe is found only in the small-diameter pipe used for field drains and around building foundations. The glazed pipe is used for sewers and other lines where the flow is caused by gravity without pressure. Most units will come with the gasket cast in the bell or spigot. Those without the gasket will have the joints sealed by placing jute and hot tar or jute and cement mortar in them.

Clay pipe comes in short lengths, usually between 3 or 4 ft long. Some lengths will be warped slightly, and if so, the warp should be laid horizontally. If the warp is laid vertically, the pipeline will have high and low places. Some engineers will require that warped pipe be discarded.

Clay pipe must be handled carefully, as it will chip easily and stones or other objects falling into the trench can break it. Bell holes should be used so the pipe will lay on its full length, not just on the bell. (See Section 3.14.)

4.28. Clearing (Site)

Most construction sites require some clearing before work on the project can proceed. The clearing may involve only the removal of sod, bushes, and trees, or it may include buildings, roadways, underground structures, water and sewer lines, and anything else in the way. You must understand what is to be removed, what materials are to be preserved, if any, and how such materials are to be disposed of. Topsoil is usually stored on the site and reused as directed. Some trees and other landscape items may require

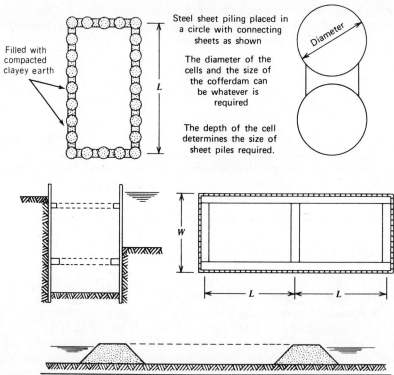

FIGURE 4.14. (top) Cofferdams. Cofferdams can be made in almost any shape and of various materials. A circular layout of steel sheet piling is used mostly for large foundations such as on a bridge pier. The circles and the areas between them are filled with earth. The diameter of the cells can be whatever size is required. The depth depends on the depth of the water and the water-bearing earth that must be penetrated to reach bedrock. The sheeting and bracing type of a cofferdam, similar to trench sheeting, can be as deep as the length of sheeting available or double that if a second stage is installed inside the upper section. (bottom) Cofferdams of a low height are made of earth dikes that surround the area to be excavated. Such dams must be well founded to prevent undermining when water seeps through. The earth must be well compacted and the slopes must be flat. When the water is deep, and especially if it is moving, the cofferdams should be designed and the installation inspected by an engineer.

protection, and if so, the specifications will indicate which. (See Section 21.25.)

When clearing, care must be taken to avoid disturbing property monuments. If any are found, your superior should be advised so orders can be issued to provide protection. The replacement of property monuments can be costly, depending on how much surveying is required.

4.29. Cofferdams

A cofferdam may be constructed of timber, steel sheet piling, earth, or a combination of these. The smaller ones are usually of timber, similar to the sheeting and bracing in a trench, except that they will be closed on all four sides. The larger, more complicated ones will be of sheet piling in two rows with clayey earth in between. (See Figures 4.14 and 4.15.)

Cofferdams are used for excavating in river or lake bottoms, in swampy areas, or where high groundwater may cause the banks to collapse. They must be founded on earth stable enough to hold the load and tight enough to prevent groundwater from seeping underneath. The construction of the dam must be strong enough to resist the pressure as the excavation proceeds downward. If the cofferdam is deep, the joints may have to be caulked with jute. (See Section 11.11.)

On large projects, the cofferdam may be constructed to be wide enough to provide room for materials and equipment. Some, such as those in a river, will be just large enough to build the structure, and all materials and equipment will be on barges alongside. Some projects will have a narrow-gage railway trestle running out from the shore. This method is particularly useful for multispan bridges.

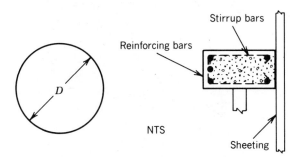

FIGURE 4.15. Reinforced concrete ring for cofferdam (should be engineer designed). The diameter D can be almost any length. They have been built up to 60 ft. The sheeting can be timber or steel. The ring can be left in place or demolished as required; however, they are costly to break up. The spacing of the rings will depend on the loads. The rings are poured in place and require support columns of timber, steel pipe, concrete, or H-columns.

In areas where the limits of the rise and fall of the water surface are known, the cofferdam will be built high enough to prevent flooding. In large rivers subject to high flooding, most contractors will arrange for advance warning of floods and will remove all equipment and let the cofferdam flood. Such flooding will leave a layer of mud in the dam, which will have to be removed before resuming operations. An easy way to remove mud is to pump the water down, place a mud pump at the lowest level, and hose down the mud with a strong jet of water. The mud will become suspended in the water and be easily pumped out.

4.30. Come-Alongs

A come-along is a device for pulling together such items as heavy pipe, units of structural frame, or piling, or for moving heavy equipment into

FIGURE 4.16. A closeup of a section of 30-in. water main being installed. The crane supports the pipe just off the bedding while the come-along pulls the joint tight. The worker pointing is instructing the crane operator to hold the pipe tight in the joint while the other worker tamps bedding up the side of the pipe to hold it in place.

place. The device is composed of two lengths of chain, each with a hook on one end. The other ends are attached to a ratchet, which pulls the chains together. (See Figure 4.16.)

4.31. Comminutors

A comminutor is a machine used in sewage treatment plants to shred or chop up the solids in the sewage. The unit will be placed just after the grit chamber in the line of flow. The size and shape of the flow channel will be shown on the plans. The unit will have a vertical rotor with steel blades having sharp cutting edges. The rotor will swing back and forth through an arc of about 120 deg. The blades will clear a steel plate by a fraction of an inch, the exact clearance being specified for the kind of sewage received. The clearance can be adjusted by moving the blades.

Sometimes an object too tough to be cut up will clog the unit, causing the sewage to back up. A second channel will allow the sewage to bypass it through another comminutor or over a weir.

The comminutor must be set to be truly vertical, and all the steel blades must be at the same distance from the plate. The unit should swing free for the full arc. It is best to operate it through a couple of full swings to be sure there is no binding.

4.32. Compacting

No structure is any better than the subgrade on which it rests, so unless the subgrade is bedrock, some compacting will be required.

The foreman can assume that the design engineer has determined the bearing value of the soil and has designed the footings accordingly. If the concrete for the footing is poured the same day the excavation is made, you can be reasonably sure that the soil will sustain the load involved. However, if the excavation has been left open for some time, and particularly if it has rained, the top 1 or 2 in. of soil will be softened and will have to be removed or compacted. The engineer or inspector will determine what is to be done. If the soil is too damp, it cannot be compacted and will have to be removed. The void left may be filled with lining (see Section 13.18) or it may be required that additional concrete be placed to make up for the extra depth.

If the soil is dry enough or if lining is used, the compacting can be done with a hand tamper for a small area or by a mechanical tamper for a large area. In either case, the compacting must be uniform over the area so the density will be the same, thus avoiding uneven settlement. (See Figures 4.17–4.19.)

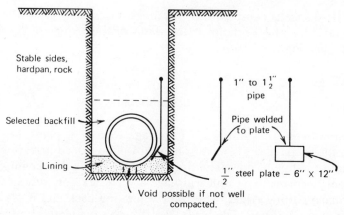

Stable sides, hardpan, rock

Selected backfill

Lining

Void possible if not well compacted.

1″ to 1½″ pipe

Pipe welded to plate

½″ steel plate — 6″ × 12″

FIGURE 4.17. Compacting iron used for lining under a pipe. Notice that the compacting iron is bent to get under the pipe. This is important because a void can develop that may allow water to seep in and soften the lining, thus allowing the pipe to settle.

FIGURE 4.18. A worker using a tamping iron. Notice the yee for the house connection and the batter board spanning the trench. The mason's line has been set to one side temporarily to allow the pipe to be lowered into the trench.

FIGURE 4.19. A closer view of a workman using a tamping iron. No lining is being used here because selected material from the excavation is available.

The hand tamper, which is a cast-iron plate from 6–12 in.^{2} has a vertical handle by which it is raised and banged down onto the soil and requires a worker with strong back and shoulder muscles.

As the figures show, there are several types of mechanical tampers, powered by gasoline motor, compressed air, or electricity. These units compact by vibration. If the area is large enough, rollers can be used. The smaller ones depend on vibration, and the larger ones depend on weight and/or vibration.

In any case, the compacting should continue until the proper density is obtained (usually the specifications require 95%). Each pass of the tamper should overlap the previous run by 50% to ensure an even density.

Some projects require that density tests be made by laboratory personnel. In the absence of this, you can be reasonably sure of 95% compaction if the tamper leaves little or no ridge as it proceeds over the surface and if there is no movement in front of a roller.

Vibration should not be used on soil of high clay and/or fine silt content. Rollers are best for these. The vibration works fine on soils of high sand and gravel content or on crushed stone.

When deep fill or an embankment is placed, the specifications will usually indicate the maximum depth of each lift, and each lift must be properly compacted before the next lift is placed. Limits of 6–8 in. per lift are most common, with 12 in. or more allowed if heavy equipment is used for compacting.

Compacting the subgrade in a trench will be done by a hand tamper (for small jobs), by a vibrator, or by a small vibro-roller. For building foundations, the vibrator or vibro-roller is used, and the 10-ton roller is used if the area is large enough. The hand tamper is used in the hard-to-get-at corners.

On a project with extensive embankment work, the specifications will require material from an approved source, which will be reasonably well-graded run-of-bank gravel with enough clay and silt content for good binding. The material will be laid in about 8-in. layers and compacted first by the sheepsfoot roller and then by the rubber-tired roller. The sheepsfoot is used first, until the feet penetrate only a couple of inches, then the rubber-tired roller is used with the sheepsfoot, and finally the rubber-tired roller alone is used.

There is another compactor, the "Barco," which has a 12–18 in. steel plate with a steel frame containing a slow-firing gasoline engine. As the engine fires, the unit jumps about 15 in. in the air and compacts as it falls back to earth. The unit has handlebars similar to those on a bicycle and requires a husky operator, as it must be kept vertical as the engine fires and the unit falls back to earth. It is somewhat difficult to obtain uniform compaction with these units, but it can be done with a capable operator.

Should you be on the job where soil technicians are not available, you can come close to 95% compaction by making the following observations. With a 10-ton steel-wheeled roller, the earth will roll up in a slight wave in front of the roller. As the compacting proceeds, the movement in front of the roller lessens, until at about 95% compaction there is very little, if any, movement. With a vibro-tamper, the ridges made by the vibrating plate will be some, but very shallow, ridges. With the sheepsfoot roller, the "feet" will sink well into the earth at first, and as the compacting continues, the penetration will lessen, until at 95% there will be very little. The results will vary with the composition of the soil, but with the absence of laboratory testing, the above methods will give very close results.

4.33. Complaints

Complaints by the neighbors or the public should be handled by an official of the contractor. (See Section 17.45.)

Complaints by one of your crew should be handled by you if possible. Listen to what the worker has to say, and if you can alleviate the problem, do so. Unpleasant working conditions, such as stormy weather, heat, or cold, will be experienced, and there is not much you can do about it. Explain that these conditions happen in construction, and make this work as comfortable as possible. Some good workers are chronic complainers and can disturb the other members of the crew. Some just like to "blow off steam" and are then okay for the rest of the day. You must learn who these workers are and let them "get it out of their system" as long as they do not disrupt the others. If you cannot get such a worker to mend his ways, you should recommend discharge. Keep a record of such problems in your diary.

4.34. Completion of the Work

As the completion of the work nears, the contractor will want to have all items of the project in condition to be accepted by the engineer so that the final payment will be prompt. The engineer will compile a list, called a checklist, of items he will want corrected before he will issue a certificate of completion. If you are given a list of items to correct, first be sure you know what is to be done to make them acceptable, and then do them correctly as quickly as possible.

A good foreman will try to do his work correctly the first time and keep the work cleaned up as he progresses in order to keep the checklist as short as possible.

4.35. Concrete (Portland Cement)

Concrete is made of Portland cement, sand, and crushed stone (or gravel) mixed with water. The sand is called "fine aggregate," and the crushed stone is called "coarse aggregate." The basic mix is 1–2–4, with one part cement, two parts sand, and four parts stone by volume. The amount of water depends on the strength of concrete required, and will usually be $3\frac{1}{2}$–5 gal/ft^3 of cement (one sack). The exact mix and quantity of water will be specified. As noted, the amount of water is critical to the strength of the concrete; it also affects the ease in placing. Some contractors like to add enough water to make a mix that will flow readily into place with little effort. This procedure will reduce the strength of the concrete, and most inspectors will not allow it. Usually, a plastic mix is specified. Such a mix will look too dry as it comes down the chute, but it will readily flow into place if properly vibrated. Too dry a mix is just as bad as a wet one, as the concrete will not flow properly into the corners of the forms or fully around the reinforcing steel and will not provide a good bond when vibrated. (See Section 7.22.)

Most concrete today is furnished by Redi-mix plants, which deliver the proper grade of concrete in 5–15-yd^3 mixer trucks. Some large projects, such as bridges or dams, will have on-site mixing in plants similar to the Redi-mix plants. In either case, a modern plant will be computer controlled, with the aggregate placed in bins above the mixer and released in measured quantities. The mixing time will be controlled, and if the concrete is to be delivered off site, water will not be added until that particular batch is needed on the job. The foreman will signal the truck driver, and he will add the water and mix the concrete in the required time. If the mixing plant is on the job site, the water will be added with the aggregate, and the completed mix will be delivered to the spot needed on the site.

The off-site Redi-mix plant will have an inspector who will certify that the mix is as ordered, and she will sign or initial the delivery slip, which the truck driver will give to you. You should make it a point to collect the delivery slips as they arrive on site and not wait to pick them up in a bunch. Check to see that the slip has been signed or initialed, and if it has not, do not accept the concrete unless ordered to do so by your superior.

For efficient handling of concrete, your crew should have rubber boots, square-point shovels, a hoe to pull the concrete out of the chute, and at least one vibrator. The size of your crew will depend on how fast the concrete is being delivered. An average crew will have a worker to position the chute or bucket, two to spread the concrete, and one on the vibrator. Sometimes the foreman handles the chute or bucket. If the pour is into narrow forms, the spreaders will handle the tremie. (See Section 21.27.)

When pouring into forms, you must keep a close check on alignment to see that the forms stay straight and do not bulge or get out of plumb. Carpenters should be available to correct any movement or failure of the forms or falsework. Any leaking through cracks in the forms should be corrected at once, as mortar leaks can cause honeycomb at that point. (See Section 9.9.)

Most plans will call for chamfer strips for all exposed corners. (See Section 7.22.) Without the strips, a sharp edge will be easily damaged, leaving exposed aggregate, which is very unsightly.

You should observe the concrete as it is delivered, looking for stone pockets, which indicate poor mixing, or for clay balls, which indicate dirty aggregate. In either case, delivery of the concrete should be stopped until the problem is corrected. A little time spent to ensure good concrete can save great expense in repairing the concrete after it has set.

After a few years of working with concrete, you get to recognize good and bad mixes by watching them flow from the mixer. A test which could be very useful until you develop your own judgment is to fill a 12-qt metal pail about 3/4 full of the mix as it comes from the mixer. Dump the pail quickly upside down on a smooth surface and then quickly raise the pail, leaving the concrete mass on the surface. The mass should settle slightly and the edges should curl under: it should not be spread out thin, nor should it stand up with straight sides. A tap on the top with a shovel should cause the mass to slump further. Such concrete may look too dry as it first comes from the mixer, but it can easily be worked around the reinforcing bars and into the corners of the forms with a vibrator.

Figure 4.20. shows how the mass will act after the pail has been withdrawn. In Figure 4.20, *A* is too dry, *B* too wet, and *C* is acceptable. A few small cracks will appear along the top edge of the mass, and if these cracks are more than 1/8 in. wide, a small amount of water should be added.

Vibration is a very important part of concrete pouring. (See Section 23.6.)

Before the finishing of any concrete slab starts, the coarse aggregate should be forced down below the surface. At least ¼ in. On small jobs, this can be done by tapping the surface with a screed board or the back of a square-point shovel. On larger jobs, it will be done with a tool sometimes called a "jitterbug." (See Section 11.4.)

The finishing of concrete slabs will be done by concrete finishers especially trained for that work. After the concrete has been screeded to the proper elevation by the concrete-placing crew, the finishers will trowel the surface smooth with steel, wood, or cork trowels, depending on the type of surface required. The work will be done by hand or power tools, depending on the size of the job.

There are several things to remember when pouring concrete. The concrete should be thoroughly mixed—not too wet or too dry, but plastic. When pouring on the ground, the subgrade should be ready for the pour, with no frost or mud. When pouring into forms, the forms should be free of debris, snow, or ice, and the concrete should be thoroughly vibrated into all corners of the forms and around all reinforcing steel. When pouring into deep columns or wall forms, gain water may accumulate, and if so, it must be drained off by drilling a hole in the forms, letting the water drain out, and then plugging the hole. When pouring deep concrete, add some

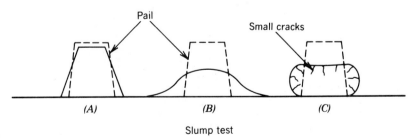

Slump test

FIGURE 4.20. A method of checking the plasticity of concrete when a slump test is not available.

FIGURE 4.21. A Redi-mix concrete truck. The truck is discharging into a concrete bucket, which will be lifted by crane to the points of discharge. This truck holds 10 yd³ of concrete.

FIGURE 4.22. A portion of the roof of one of the circular tanks. The 6 in. concrete roof slab has reinforcing steel in top and bottom. The concrete is placed rather dry, starting at the bottom of the dome and working upward. The concrete must be placed symmetrically around the tank so as to load the forms evenly. Because the rather dry concrete is porous, the surface is sealed with an application of bituminous mastic applied hot after the concrete has dried out. The concrete tank walls are faced with brick.

FIGURE 4.23. Base slab for a circular concrete tank. Concrete starting at the center is about half poured. Note the well points in the foreground.

on top to allow for settlement, and the finisher will remove any excess. When pouring T-beams, pour the vertical stem first and allow the concrete to settle before pouring the slab. (See Figures 4.21–4.23.)

4.36. Concrete Cylinders

The inspector or laboratory personnel will cast concrete cylinders at the job site for testing the strength of the concrete. The cylinders are 6 in. in diameter and 12 in. high and are cast in metal or fiber forms. The concrete will be taken from a batch as it is being poured.

The cylinders are usually left on the site for 24 h and then are taken to the laboratory for storage until it is time for testing. There will be two or three cylinders cast at a time, to be broken in 7, 14, and 28 days, as required by the specifications. When only two are cast, they will be broken in 14 and 18 days. The early breaks are to predict the final strength and allow earlier removal of some forms.

The cylinders must not be disturbed while on the job site overnight. (See Section 20.32.) All test cylinders must be protected from the weather just as the concrete in the forms.

The contractor may cast cylinders for his own use in checking the quality of the concrete, but forms can be removed only on approval by the engineer or his inspector.

4.37. Concrete Pipe

Concrete pipe may be plain or reinforced. The plain ones are used for shallow depths. The pipe will carry the flow of almost any liquid, except those containing strong acids or chemicals. The pipe sections are heavy and, except for the very small sizes, will require a crane for installation. A cable, rope, or chain sling will be used, except that the very large units may have a hole at the center, through which a cable loop will be inserted and a short rod will be placed in the loop.

The reinforced pipe will have a core of either a steel pipe or reinforcing bars. Most of them will have steel plates at the bell and spigot, with a groove for the gasket. The plain pipe may have a gasket, but usually the joint will be sealed with jute and mortar. (See Section 17.17.)

4.38. Construction Contracts

A construction contract is a legal document covering an agreement between a contractor and the owner. The owner may be a municipality such as a town, a country, a city or the federal government, or it may be a private corporation. The contract documents include the written agreement on what is to be done, bonds, insurance policies, the plans and specifications, and the costs (what and how the contractor is to be paid). The contract spells out the obligations and limitations on the contractor and the owner.

4.39. Construction Joints

Construction joints are placed in concrete slabs at points where it is expected that a crack will occur due to shrinking as the concrete sets and dries out. Without the joint, the concrete would crack in an irregular and unsightly line. Such joints are placed at the end of a day's run, again to make a neat line. The plans will show where construction joints will be required and the design to be used. Some will have bolts, as shown in Figure 4.24.

Contraction joints are sometimes specified. These are made by sawing a groove across the slab at the points indicated. Such joints are made with a carborundum saw, making a groove about ¼ in. in width and 1 in. or more in depth. This groove will make a point of weakness where the slab will crack as it shrinks, again making a straight, neat crack. These grooves should not be deep enough to expose any reinforcing steel.

4.40. Construction Managers

A construction manager is used instead of a resident engineer on some large projects, although some jobs will have both. The manager is more

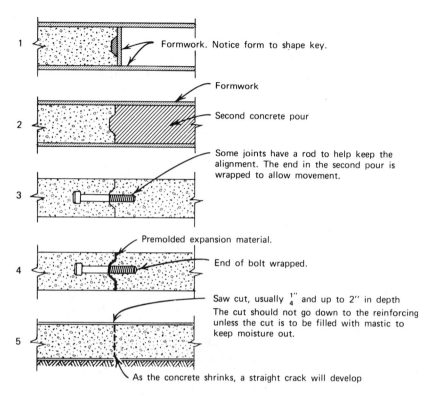

1   Formwork. Notice form to shape key.

   Formwork

2   Second concrete pour

3   Some joints have a rod to help keep the alignment. The end in the second pour is wrapped to allow movement.

   Premolded expansion material.

4   End of bolt wrapped.

   Saw cut, usually $\frac{1''}{4}$ and up to 2'' in depth
The cut should not go down to the reinforcing unless the cut is to be filled with mastic to keep moisture out.

5   As the concrete shrinks, a straight crack will develop

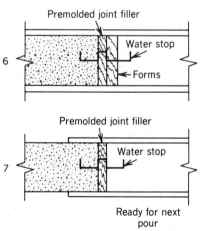

Premolded joint filler

Water stop

6   Forms

Premolded joint filler

Water stop

7

Ready for next pour

FIGURE 4.24. Construction joints. (1) Form at end of pour to make a key for the next pour. (2) Form removed at key, and second pour in place. (3) Some joints have a rod to keep the slabs in alignment. Before the second pour, the end of the rod is wrapped with paper or cloth to prevent bond and allow movement. (4) Same joint as at (3) except premolded expansion material has been added before the second pour to allow for expansion. (5) A ¼-in. wide saw cut up to 2 in. in depth, but not deep enough to expose the reinforcing steel. A bituminous mastic is poured into the saw cut to keep moisture and dirt out. This cut causes the concrete slab to crack at this point in a straight line and not an irregular line, which would be the case without the saw cut. All joints prevent cracking. Only joints with expansion material allow expansion. (6) shows method of placing forms when a metal or plastic waterstop is used with a premoulded joint filler. (7) same as sketch (6) except the end form has been removed ready for the next concrete pour.

than a resident engineer in that he will usually be hired at the conception of the project and assist in the original planning, assist in the hiring of the engineer and/or the architect, and participate in the development of the contract documents and the selection of the contractor. During construction, he will coordinate the arrival of materials and equipment at the job site.

4.41. Construction Methods (General)

As noted in the Introduction, each contractor will have his own methods of operating, depending on his experience and the kind of equipment he chooses to use. Methods are discussed under the various items of work.

Some general rules apply to all projects, such as those of safety, of causing the least inconvenience to the public, and of causing minimum damage to the site to reduce the cost of restoration. The quality of the work should be such as to avoid the expense of replacing rejected work. Federal, state, and local laws must be obeyed. There should be a minimum of damage to or trespassing on adjoining property to avoid claims by the owner.

Booklets can be obtained from the Portland Cement Association and the American Concrete Institute on concrete work, from the American Institute of Steel Construction for structural-steel work, and from the various pipe manufacturers on pipe laying. Information on timber and wood can be obtained from the United States Department of Agriculture and from the National Lumber Manufacturers Association. The addresses for these groups can be obtained from your local library.

Noise, dust, and traffic interference affect the public relations of the contractor. How much time your crew spends on these problems will depend on instructions from your superior. Some contractors are concerned about their public relations, while others ignore them until instructed by the inspector or engineer.

4.42. Construction Methods (Summer)

Summer construction is much less difficult than winter; however, there are problems. Structural steel can become so hot that if a worker touches it, his reflex action could cause him to slip or fall. If the steel is exposed to the sun for long periods, it will expand and cause misalignment of the holes at connections. This may require that the steel erection be done early and late in the day or that the steel be hosed down with cool water. Hot, windy days can cause concrete to dry out before it has properly set, causing cracks, particularly in concrete slabs. In hot weather, the forms or subgrade should be dampened just before the concrete is placed. The surface of concrete slabs should be sprayed with moisture or a surface sealant. When placing an embankment, it may be necessary to sprinkle the area to maintain the optimum moisture. Dust is always a problem in hot weather, requiring sprinkling or spreading a dust retardant. If the project is not big enough to have a sprinkler truck, one can be improvised with a 55-gal drum of water connected to a 1-in. pipe with small holes it drilled at about 4-in. centers. The drum is then mounted on the back of a pickup truck.

4.43. Construction Methods (Winter)

The contractor's costs rise considerably in winter. Cold-weather clothing restricts the free movement of the workers, and the equipment is harder to start and operate. The cold weather requires special protection for concrete and masonry work. It will stop bituminous-concrete work. Earthwork is restricted if frost is present. Daylight hours are shortened. Snow can restrict the free movement of workers and equipment. Forms and reinforcing bars must be kept free of ice, snow, or frost. Poured concrete must be kept above 50° F until it hardens and dries out. If masonry work must proceed, it should be enclosed in a framework covered with plastic sheets, and the enclosed area should be heated. Plastic sheets are better than tarpaulins because they let in the daylight. (See Figs. 4.25 and 4.26.)

Pipe laying is more difficult because of the necessity of keeping frost out of the subgrade and backfill. Work gloves make it more difficult to handle the pipe, lubricate and install the gaskets, and handle the tools.

Snow and ice get on the scaffolding plans and must be constantly removed to provide safe footing. Sand is often spread on the planks after the snow and ice are removed.

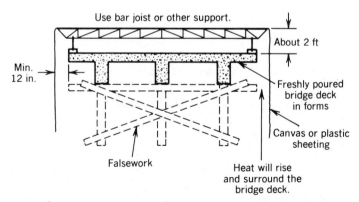

FIGURE 4.25. A canvas or plastic sheet draped over the structure as shown and extended down to the ground or water surface will envelop the structure with enough heat to prevent freezing. A space heater such as shown in Fig. 4.26 can supply enough heat for about 20 ft. of length. The bottom of the sheets should be tied down over land and weighted down over water.

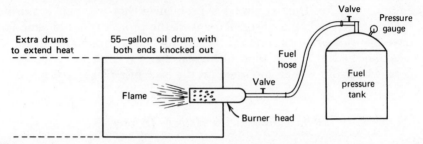

FIGURE 4.26. Space heater used during winter construction. A row of 55 gal steel or corrugated metal drums distribute the heat evenly. There can be as many as 10 drums for one 4-in. burner. There should be at least one drum to protect the workers from the flame.

Winter headgear will restrict sight and hearing. Gloves become wet and slippery, causing workers to drop tools or loose their grip as they move about. Feet get cold, making walking unsteady.

If concrete is mixed on the site, the frosty aggregate must be heated long enough to be warm clear through. Aggregate that has been heated quickly may feel warm to the touch and still have frost inside. This frost will come out after the concrete has been poured and will prevent proper bonding with the mortar.

Control stakes placed before construction can be displaced by frost heave in the ground.

4.44. Construction Specifications Institute (CSI)

The initials CSI stand for Construction Specifications Institute, a recognized authority on the format and wording of construction specifications.

4.45. Contours (Contour Lines)

Contour lines on a map indicate the elevation and slope of the earth's surface. By noting the difference in elevation between two lines and the horizontal distance between them, you can obtain the slope of the ground. If you are involved with grading the site for landscaping, you will be involved with contour lines which shape the earth's surface to provide the drainage required.

On a job of any size, the contour lines will be indicated by surveyor's stakes. The stakes will indicate the meandering line of the contour, and all points on a line will be at the same elevation. When contour lines are close together, they indicate a steep slope. The farther apart they are, the flatter the slope will be. Referring to Figure 4.27, the numbers 135 on a line

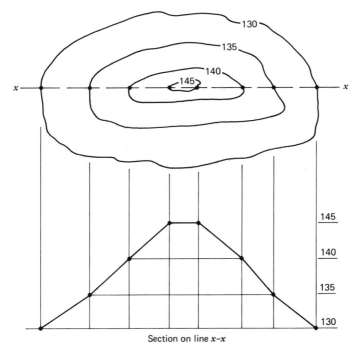

Section on line *x–x*

FIGURE 4.27. Contours.

indicate that all points on that line are at 135 ft above sea level or at some other specified datum. Note that the numbers run from 130 on the outside line to 145 on the inside line, indicating a rise. If the numbers were reversed, with 145 on the outside and 130 on the inside, they would indicate a depression rather than a hill.

4.46. Contraction Joints (See Section 4.39)

4.47. Control Panels

Control panels provide a central location where the off–on switches for the various units of equipment and the light circuits throughout the plant can be grouped for ease in observing. Such panel will usually have warning lights to indicate if the power is on or off and if there is a malfunction. Some have alarms which sound when there is a malfunction or if the power goes off. Some have a locking system so that any unit can be locked in the off position while repairmen are working on the unit.

The smaller units are contained within metal cabinets, and the wiring is done from the front. The larger units will have a space or room behind the panel where the wiring is done. If a considerable amount of power is required, there will be heavy copper rods along the top (bus-bars), with wire connections to the various controls. The installation of a large control panel takes several days, so the area should be well barricaded, and only authorized persons should be allowed in the area. All power should be shut off at night, as children may explore the area after the workers have left.

Usually, a control panel will be inspected and checked by an engineer. If you are involved, do not handle the wiring; leave it up to the electricians to do their job, and you be concerned with seeing that the start–stop controls start and stop the proper pieces of equipment; if not, call the electrician.

4.48. Conversion Tables

The government is planning to convert to the metric system, so you should be familiar with the various units. Complete conversion tables can be purchased from most bookstores. You should have a set and become familiar with the units for volume, length, weight, and weather. (See Appendix I.)

4.49. Corbels

Figures 4.28 and 4.29 indicate what a corbel is. A corbel can be made of brick, concrete, blocks, or poured concrete. They are similar to cantilevers, except their span is very much shorter and they do not require reinforcement. Masons will lay the corbels, except when they are poured concrete, in which case carpenters will shape the forms as required.

4.50. Cores

Some contracts require that "core samples" be taken to check the quality of some materials, such as the subgrade, a concrete slab, a weld in a steel joint, or any material requiring a check on its interior. The core will be obtained by drilling with a hollow drill that retains a portion of the material for examination or testing.

4.51. Corrugated Metal Pipe

Corrugated metal pipe is used for roadway culverts, storm sewers, and other drainage lines. The pipe can be painted, galvanized, or bitumen coated. The metal can be steel or aluminum. The lengths will be connected by a

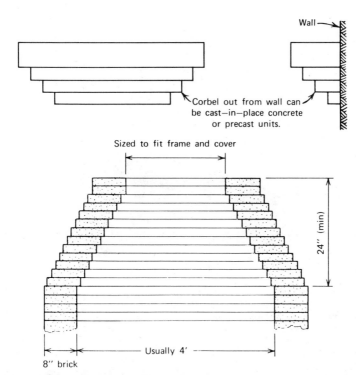

Wall

Corbel out from wall can
be cast—in—place concrete
or precast units.

Sized to fit frame and cover

24" (min)

Usually 4'

8" brick

FIGURE 4.28. Corbel. The illustration (top) shows a corbel of stone or brick forming a ledge on the wall of a building and (bottom) the corbel at the top of a brick or concrete block manhole. The outside of the manhole is usually parged with cement mortar, although that is not required in areas having no groundwater.

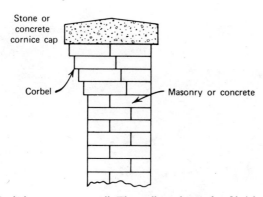

Stone or
concrete
cornice cap

Corbel

Masonry or concrete

FIGURE 4.29. Corbel on a parapet wall. The wall can be made of brick or concrete blocks.

FIGURE 4.30. A section of a 24-in. corrugated metal pipe treated with a bituminous coating for use as a storm sewer.

FIGURE 4.31. The corrugated metal collars used to connect the sections of pipe. Note that the corrugations are spiral, except at the ends where they are at right angles to fit the corrugations in the collar.

collar, with corrugations to fit the ends of the pipe. The collar must fit the end corrugations exactly, and the bolts must be tightened as tightly as possible without stripping the threads.

Corrugated metal pipe is not as strong as most other pipes, so it should be well supported by fully compacted backfill under it and up around the sides.

Connections that do not require gaskets are preferred, but some do require them, and if you are involved with one that does, the ends of the pipe sections should be tight together, as sand and other grit in fast-moving storm water can quickly erode the gaskets, causing the joint to leak. (See Figures 4.30–4.34.)

FIGURE 4.32. A section of 24-in. bitumen-treated, corrugated metal pipe being placed for a storm sewer. Note the cable attached to the backhoe bucket used to place the section.

FIGURE 4.33. Worker placing lining on the sides of the pipe.

4.52. Cost-Plus Work

Cost-plus work is usually much more expensive than unit-price or lump-sum work unless the contractor is very conscientious. The cost-plus means that the contractor does the work, keeping the costs of each item, and then adds a percentage of the costs as his overhead and profit. The more he spends, the more he makes, so quite often there is no incentive to keep costs under control.

Cost-plus is also used when a job must be done quickly, with no time to study the job and arrive at a figure for the work. Such jobs will usually be well inspected to keep track of the costs.

FIGURE 4.34. Section in place with lining under and on each side. Note that the backfilling has been completed in the background.

4.53. Couplings

A coupling is a device for connecting together units of equipment such as railway cars, a trailer to a tractor–truck, and a bulldozer to earth-moving equipment. It is used any place where two units of equipment must be connected.

The term is also used to mean the threaded connection of metal pipe.

4.54. Courts (See Section 24.23.)

4.55. *Cracks*

All concrete shrinks in volume as it sets and dries out. The resultant cracking is controlled by properly sized and placed reinforcing bars and construction joints or expansion joints. The plans will show how each of these should be placed, and you must see that the details are followed. There are several ways to protect concrete in hot or cold weather, so you must learn your employer's method and follow it.

4.56. *Cradles*

Cradles are used when pipe is being laid on an unstable subgrade or on a concrete mat. (See Figure 4.35.) The pipe will be laid to line and grade, supported on concrete blocks so the concrete can flow under and up the sides of the pipe. The cradle cushions the pipe, spreading the load on its support and increasing its resistance to the load of earth backfill above it. The plans will indicate the size and spacing of the concrete support blocks.

4.57. *Cranes*

Cranes may be part of the permanent equipment of a commercial or industrial project, and if so, they will be installed by factory crews.

   The contractor will have one or more cranes for erecting structural steel, placing concrete by bucket, lowering pipe into a trench, excavating by dragline or clamshell, and many other jobs on the site.

   If your crew has a crane to work with, you and the operator must know the capabilities and limitations of the machine. A plate on the side

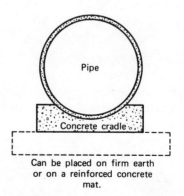

FIGURE 4.35. A concrete cradle with or without a mat.

FIGURE 4.36. A 60-ton crane. The crane on a truck body with outriggers can lift 60 tons when the boom is at an angle with the vertical of 45 deg or less.

of the machine should give information on the number of tons that can be lifted at various angles of the boom. You must be sure that the boom does not swing anywhere near overhead power cables. The earth must be stable enough to hold the machine, or timber mats may be required.

If you are using a crane in or near traffic lanes, you should have a flagman to warn when the rear end of the crane may swing into the traffic pattern. (See Figure 4.36.)

4.58. Creosote

Special care must be taken when handling creosoted timber. Contact with the creosote can irritate the skin of some workers, and in hot weather a vapor given off by the creosote can also irritate the skin, especially if the pores are open due to sweating. All workers should wear long-sleeved shirts and gloves when working with creosoted timber.

The timber is treated with creosote to preserve it, so it should penetrate the full depth of the fiber; however, sometimes the center of a timber is not treated. So when a timber is cut, check to see that all fibers are treated, and if not, hand paint the untreated area with brushed-on creosote. Make several applications so the fiber will absorb as much as possible.

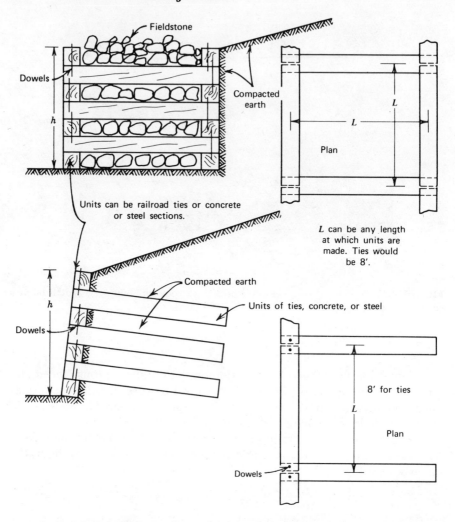

FIGURE 4.37. Cribbing. Cribbing can be a long series of bins filled with rock or stones as shown (top), or it can be on the face of an exposed bank with the ties set in trenches cut back into the earth. The earth is then backfilled around the ties and compacted to hold them secure. Cribbing can be any reasonable length and height.

4.59. Cribbing

Cribbing is a framework, usually of timber, although it can be of metal or concrete. It is used to contain a fill or embankment or as a temporary support under a structure. (See Figure 4.37.)

4.60. Critical Path Method (CPM)

Many large projects are now using the CPM to control the progress of the work.

The CPM is a bar or graph chart on which all the various operations are charted in the sequence in which they will be undertaken on the project. The chart covers the initial planning, preliminary designs, cost estimates, final designs, all field operations required to complete the job, and the times various units of the project will be required on the site.

The CPM is particularly useful when some units of equipment are delayed. The chart can be studied and the schedule of operations changed to keep the work moving.

4.61. Cross Connections

In plants with considerable piping, a cross connection can inadvertently be made which will connect the potable water supply to an on-site supply from wells or other sources and cause contamination. Sometimes potable and unpotable water are discharged into the same tank, and if so, there should be check valves to prevent backup.

If you are involved with installing such piping, be very sure all connections are made as shown on the plans and that pipes carrying potable water are never connected to those carrying unpotable water, unless there are check valves to control the flow.

4.62. Cross Sections

A cross section on a plan shows what the object would look like if it were cut vertically at the point indicated. (See Figures 4.38 and 24.22.) The first figure shows how an excavation would look if cut through the earth. Such sections are used to calculate the volume of earth moved. The same thing

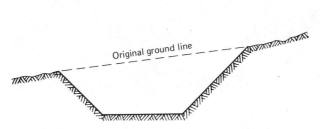

FIGURE 4.38. Cross section of roadway excavation.

can be done with an embankment. The upper part of the second figure shows a section through a frame structure from which the carpenter can tell the sizes and lengths of the various pieces.

4.63. Crowns

A crown, or high point at the center of a roadway, allows the rainwater to run off quickly. This is particularly important on secondary roads, which have very little surface treatment. The water on the surface softens the material, and traffic will break it up, causing potholes. The crown slope should be at least ¼ in./ft of width and should be more on lower-grade roads.

4.64. Cubic Feet Per Second (CFS)

Cubic feet per second (CFS) is a unit of measure to indicate the volume of flow of water in a stream or over a dam or weir, or the flow of water or other liquid through a pipe. CFS can be converted to gallons per second (gps) by multiplying the amount by 7.48052.

FIGURE 4.39. A four-barreled reinforced-concrete culvert under a state highway. This design is used for culverts of one or more barrels, as needed to carry a large volume of water at low depths. Note the wings, or retaining walls, on each side.

4.65. Culverts

A culvert is a drainage structure. It can be a concrete box, as shown in Figure 4.39, or it may be a pipe. Culverts are in fact small bridges. Some are used for drainage and some allow farm animals to pass under the roadway.

When corrugated metal pipe is used, it must be well backfilled. (See Section 4.51.)

A poured-concrete culvert is built like any other concrete structure. (See Sections 4.35, and 19.5.)

4.66. Curb Boxes

When a lateral pipe is run from a water main to a residence or other structure, the pipe within the street right of way will be placed by the contractor placing the main or later by the water company. A valve will be placed at the end of such a pipe, and the plumber will connect this valve to continue the line into the structure. Such a valve is called a curb box, as it will be at or near the curb.

4.67. Curbs

Curbs can be made of granite sections, concrete, or bituminous concrete. The granite sections will be cut stone 4–6 in. thick, 18–24 in. high, and

FIGURE 4.40. Installing a concrete curb. The forms on the curve have just been filled with concrete. The workmen are setting a cast-iron catch-basin frame and grate before pouring the remainder of the curb.

4–8 ft long. The sections may be set vertically or sloping outward slightly and will be on a bed of sand, fine gravel, or concrete, although they can be set directly on the earth if it is stable.

The poured-concrete curb will have forms to allow the pouring of the curb and gutter as one section. The forms will be curved to go around the corners at intersections. The concrete may be reinforced. (See Figure 4.40.)

The bituminous-concrete curb is cast in place by a special machine that forms the curb as it moves along the line. (See Figure 4.41.)

The granite and bituminous curbing must have backfill placed behind it as soon as possible to keep the curb in line.

Granite curbing will have mortar in the joint between the sections, concrete curbing will have construction joints, and bituminous curbing will be continuous.

Provisions must be made for driveways and street intersections. The plans will give the details.

Any curbing must be straight and on line and grade between curves, as a wavy curb is unsightly and indicates poor workmanship. (See Figures 4.42 and 4.43.)

FIGURE 4.41. Installing a bituminous-concrete curb. The bituminous-concrete curb is installed by use of a slip-form machine. The machine moves along a line marked on the pavement. the bituminous mix is shoveled into the hopper of the machine, where it is compacted as the machine moves along. This is an economical curb, but backfill must be put in behind it as soon as it is laid or it can be pushed out of line. Care must be taken when backfilling not to force the curb out of line. Once the curb has been carefully backfilled and allowed to set about 24 h, it serves very well.

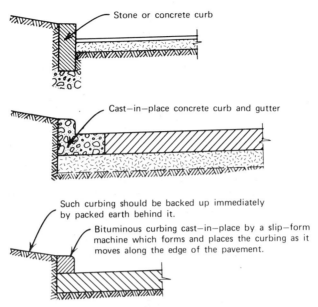

Stone or concrete curb

Cast—in—place concrete curb and gutter

Such curbing should be backed up immediately by packed earth behind it.

Bituminous curbing cast—in—place by a slip—form machine which forms and places the curbing as it moves along the edge of the pavement.

FIGURE 4.42. Three different types of curbing. (top) A precast concrete or cut-stone curb set on a gravel base before the roadway is placed. (center) A cast-in-place concrete curb set on the roadway base course. The remainder of the roadway is installed after the curb has hardened. (bottom) A cast-in-place bituminous curb set on the roadway and backfilled.

FIGURE 4.43. A newly placed concrete curb and sidewalk on a street improvement project. Note that an opening has been left for a catch-basin grating. On the left-hand half of the roadway, surfacing has been installed.

109

4.68. Curves

Curves in the alignment of roads or streets, curbs or sidewalks, or pipelines will have special stakes to indicate where they start and stop, whether they are horizontal or vertical, and the degree of slope. You must know what each stake represents and the dimensions involved. Special care must be taken in the layout and construction of curves, as it is harder to detect errors on a curve than on a straight line. Misalignment is always unsightly. If you are placing granite or bituminous curbs, they must be backfilled at once to keep them on line. Because curbs require small amounts of concrete per foot, the forms are sometimes flimsy and easily pushed out of line. Such forms should be carefully checked constantly.

4.69. Cut and Fill

Cut means to excavate on the earth's surface to change or lower the grade to accommodate the proposed construction or to change the contours to improve the surface drainage. Cut can also mean the distance below a grade stake that the top of a footing is to be constructed or the invert of a pipe is to be placed.

Fill means the opposite of cut. Fill is the placing of excavated earth or other material on the existing earth surface to raise the elevation. It also means the distance above a grade stake that the top of a footing, slab, or embankment is to be placed.

5

5.1. Daily Reports (See Section 19.3)

5.2. Damproofing (See Section 23.2)

5.3. Dams

Dam construction can range from construction of a small earth embankment to a very large concrete structure.

The subgrade on which the dam is founded is extremely important in preventing undermining or slipping. Most dams are founded on bedrock when possible. The next best thing is a well graded mixture of sand and gravel with enough clay to make the mass impervious.

The area of the site will have been examined and soil samples will have been taken by the engineer before design is undertaken, so a foreman will have very little to worry about concerning the adequacy of the site as a whole. However, you should be alert to any changes as the excavation proceeds. If changes occur, notify your superiors and/or the engineer at once.

Most dams of over a few feet in height will have a curtain wall down into the subgrade if it is not bedrock. The details will be on the plans.

Construction methods and your responsibilities are discussed under excavation, embankments, concrete forms, drainage, and so on, elsewhere in the book.

The curtain wall can be constructed by trenching down to impervious material and filling the trench with dense concrete or steel sheet piling. The plans will give the details.

Figures 5.1 and 5.2 show the two usual types of dams. (See also Figures 5.3–5.10.)

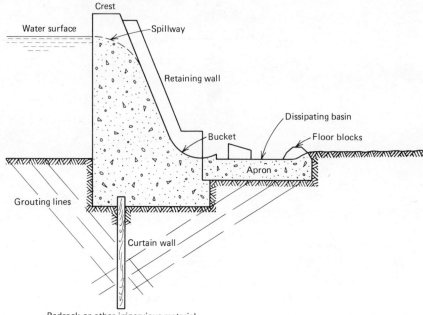

FIGURE 5.1. The general design of a concrete dam. If possible, such a dam should be founded on bedrock. If this is not so, most such dams will have a solid curtain wall reaching down to bedrock or other impervious material. If no curtain wall is used, grout under pressure may be injected. The plans and specifications will indicate the requirements. The curved area at the base of the downstream slope is called the bucket. It deflects the water upward to prevent scouring of the stream bed. Sometimes an apron is used to prevent further scouring, and if there is a large volume of water, floor blocks are placed at the end of the apron to create a dissipating basin. There will be retaining walls on each side of the spillway to confine the water to that area. The concrete spillway of an earthen dam will have some or all of these features. Floor blocks must be placed exactly as shown on the plans.

5.4. Darbies

A darby is a long, narrow hand-held trowel used to finish concrete surfaces. It may be made of wood or wood with a cork surface. It is usually about 4 in. wide and 24 in. or more long.

5.5. Deadmen

(See Section 21.15.)

A deadman is a device for anchoring a restraining cable or rod. (See Figure 21.10.) The size depends on the pull on the cable or rod and the bearing value of the soil.

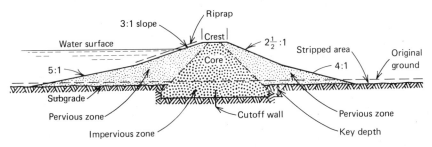

FIGURE 5.2. The general design of an earth dam with an impervious core that must be founded on a stable subgrade. The depth of the core into the subgrade depends on the type of material encountered. Before construction starts, the area to be occupied by the dam will be stripped of all vegetation, topsoil, silt, and other objectional material down to an acceptable subgrade. The stripped zone can be from a few inches to several feet in depth. A proper core is the key to a successful earth dam. It must be of the proper material, thoroughly compacted in layers and well founded on a stable subgrade. Some designers require a cutoff wall. The core will usually be a mixture of sand, gravel, and clay, with the clay content being high enough to bind the mixture and fill all the voids. The mix should be placed at optimum moisture content. The pervious zones will be run-of-bank gravel, run-of-crusher stone, crushed stone, or a mixture of these. Dumped riprap makes an excellent surface for the upstream slope. Placed riprap may be required in areas where the movement of the water could erode the slope. The downstream slope outside the spillway can be crushed stone, or it may have topsoil and seed for a grass surface.

FIGURE 5.3. The start of excavation into the canyon wall at the site of a concrete dam. The ends of the dam must be carried into the wall far enough to prevent seepage around the end, the distance depending on the type of material. The excavated material has to be disposed of off site. In this case, it was used to construct an embankment or levee upstream where a narrow valley had to be sealed to form the lake behind the dam.

FIGURE 5.4. One end of the dam has been poured, and excavation has been started on the other end. It is customary to construct a dam from both ends toward the middle to avoid disturbing the main flow of the river for as long as possible. During the construction of the middle section, the river will flow through a drainage gate installed to allow for future drainage of the lake if necessary.

FIGURE 5.5. A view from upstream showing one-half of the dam poured. Note the drainage gate opening at the riverbed. a heavy cast-iron gate will be attached to the opening and will be operated from a gatehouse on top of the dam. Note that the river has been shifted to one side by temporary embankment.

FIGURE 5.6. One section of the dam poured to its full height. The horizontal lines indicate the depths of the various pours. Note the recess in the end of the section. The dark line in the center of the recess is a rubber water stop. The next pours on this side of the section will fill the recess and envelop the water stop to prevent leaks through the construction joint. The water stop will be one continuous length and must be protected from damage as each pour is placed above the lower one. The water stop will extend to an elevation well above the surface of the lake. Note the forms in place for pouring another section of the parapet wall on top of the dam.

FIGURE 5.7. Forms in place for further pours near the top of the dam. Note the reinforcing steel dowels on the downstream face of the dam. These are for a retaining wall that will confine the flow of water over the spillway.

115

FIGURE 5.8. Forms for one of the final pours on the spillway. Note the retaining walls on each side of the spillway.

FIGURE 5.9. Grout pipes in the lower upstream face of the dam. A hose for temporary water supply has been strung along the face of the dam. These pipes extend to the subgrade below the dam and through them grout under pressure will be forced into any possible voids to prevent leakage under the dam. On completion of the grouting, the holes will be sealed with concrete.

116

FIGURE 5.10. Completed dam with water flowing over the spillway.

5.6. *Dewatering*

(See Sections 24.14 and 24.15.)

Dewatering of an area may be necessary if the groundwater is high enough to interfere with the excavation.

Problems that arise are disposal of the water, lowering the water table to an extent that dries up or reduces the water-supply wells in the area, and settlement of the earth under the footings or foundations of structures in the area.

If there is a possibility that the lowering of the groundwater could cause complaints or lawsuits, it is advisable to install test wells around the area. The simplest method of installing test wells is to jet down a well-point riser pipe, at several locations around the site, establish an elevation on the top of each pipe, and take readings of the water elevation several times a day. A heavy bolt wired to the end of a measuring tape can be lowered into the 1½ in. pipe to check the water elevation. A cap should be screwed onto the top of the riser pipe to keep things from falling or being dropped in.

Your responsibility could be to record the well elevations. If so, make it a practice to take the readings at the same time each day, and keep neat, accurate notes of the readings, as you might someday have to testify about them in court.

It should be determined what elevation of the groundwater might be critical to the wells and foundations in the area, and your superiors should be notified if the readings indicate that the elevation might be reached.

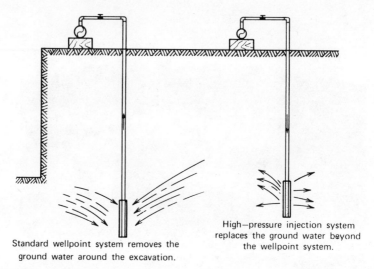

Standard wellpoint system removes the
ground water around the excavation.

High—pressure injection system
replaces the ground water beyond
the wellpoint system.

FIGURE 5.11. Recharging groundwater. The well-point system removes the groundwater from around the excavation, while the high-pressure injection system replaces the groundwater to prevent settling beyond the work area.

Should there be an indication that the groundwater lowering might affect the wells or foundations, provisions can be made to inject water back into the ground (See Section 24.14 and see Figure 5.11.)

If the area of the proposed construction has groundwater at or near the surface, or if the area is a swamp, drainage can be accomplished by digging several ditches (see French drains), digging one or more pits, and

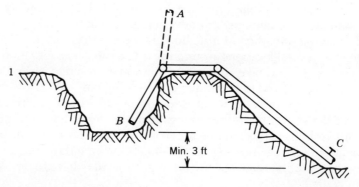

FIGURE 5.12. To fill the siphon pipe, raise the upper section to position *A*. Close valve *C*. Fill the pipe with water and plug the end with a rag. Lower the pipe section, the *B* position. Open valve *C* and pull the plug at the same time, and the siphon will start. Note the pipe joints must be airtight. If they leak, seal them with thick mud.

pumping the water away, or if there is lower ground in the vicinity, the water can be siphoned off (see Section 20.26). There is always the problem of disposing of the water without damaging other areas. (See Figure 5.12.)

Care must be taken to see that the water being removed from the site follows the course intended and does not become diverted or dammed and cause problems on adjacent lands. (See Section 17.49. for the kinds and sizes of the different pumps used for dewatering.)

5.7. Diamond-Point Drills

When drilling in very hard rock, steel-pointed drills become dull quickly, so diamonds are added to extend the useful life of the drill. Such drills will be shaped like a star drill with small diamonds along the ridges. (See Section 20.49.) Small-diameter drills will have a single point with a small diamond at the point.

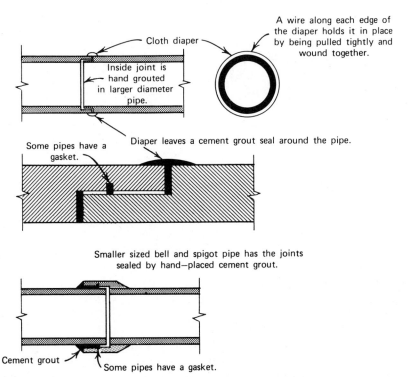

FIGURE 5.13. Diaper use with sewer pipe. A diaper of strong fabric is held in place by wire along each side. The diaper is pulled tight and the wire are wound tight. The grout is poured in at the top, making a grout seal for the pipe joint. When large pipe is used, the inside of the joint is filled by hand. When small pipe is used, the bell-and-spigot joint is filled by hand with cement grout.

5.8. Diapers (for Pipe)

Some types of concrete pipe use diapers to grout-seal the exterior void at
the joints. See Figure 5.13. The diaper, made of canvas or other fabric, will
be a strip 6–12 in. wide and long enough to go around the pipe at the
joint. There will be a wire along each long edge, by which the diaper is
secured to the pipe. A rather thin grout will be poured in at the top until
the void in the joint and the diaper is full. This will cause the diaper to
expand several inches and form a mound over the joint. The diaper is left
in place until the grout has set, and it may or may not be removed. Care
must be taken to see that the wires are pulled tight; otherwise, the grout
will leak out and leave a defective seal.

5.9. Diaries

It is desirable to keep a diary covering your actions each day and it is
recommended to anyone in charge of work. When on the ground in charge
of work, writing in a diary can be messy, so keep a notebook in your pocket
and jot down any orders you receive and from whom they are received.
Also, record any problems that arise with labor, equipment, or materials.
Develop your own shorthand, and at the end of the day transpose the notes
into the diary, thus keeping it neat and clean. The diary is a great help in
recalling situations several years later. The notes should be transferred to
the diary within 24 h, while the details are fresh in your mind. You may
have to use your diaries while a witness in court or arbitration proceedings,
and the judge will place much weight on notes which were recorded while
fresh in the mind.

Diaries should be kept at least five years, and then, if there is no
indication of problems, they can be disposed of.

5.10. Discharging Workers

Discharging workers is always an unpleasant task. Usually, discharging is
handled by higher authorities, with the foreman only recommending such
action when he deems it necessary.

Most people must work to earn a living, and they should not be
deprived of the opportunity to do so unless they are not doing the work
properly or are causing disruptions on the job.

There must be no personality clash, and a worker should be removed
only after careful consideration of all the facts.

If you find one of your workers cannot or will not follow instructions
and you have conferred with him without success, you should confer with

your superior about transferring the worker to another part of the project under another foreman or about discharging him.

A foreman must be a leader, and as such, should be able to organize his crew so they will work together effectively. However, sometimes there is a worker who just does not fit in, or does not want to, and he should be removed to keep harmony among the rest of the crew. If you are satisfied that a worker must be removed for the good of the project, do so without remorse. Some foremen get a guilt feeling if they are required to discharge a worker, no matter how justified the discharge, and this can affect their own performance for several days. Always be sure in your own mind that you are justified, and then take action and forget it.

If you find that you must recommend the discharge or transfer of one of your crew, make detailed notes in your diary of all the facts and why you thought as you did. Do not accuse a worker of anything that you are not sure he is guilty of doing.

5.11. Diving

Diving is a dangerous occupation and should not be undertaken without proper instruction by a qualified diver.

There are three classes of divers, depending on the depth and the time submerged. For shallow depths and short periods, the snorkel face mask is used. For greater depths, scuba equipment, which consists of a heavy face mask, air tanks strapped to the back, and hoses connecting them, is used. For greater depths, deep-sea diver's equipment is used. This consists of a complete rubber or waterproof fabric, metal headgear, and weights to hold the diver down. Air under pressure is forced into the headgear and suit, which allows the diver to go down to 200 ft and stay there for up to an hour. He must descend and ascend slowly. There are very strict rules governing deep-sea diving.

If you are interested in diving, attend an accredited school and learn the rules well before you attempt to supervise others.

5.12. Dolphins

Dolphins are clusters of piles, usually three, five, or seven, banded together with cable or iron straps. They are used to anchor ships offshore, in rows to guide ships into the proper channel, and as protection along a dock where the ships tie up. The piles must be driven deep to take the pressure of a large ship coming in too fast or subject to high winds. (See Figure 5.14.)

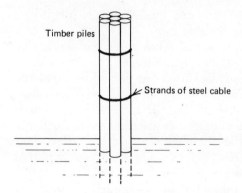

FIGURE 5.14. Dolphin.

5.13. Doors

Doors will usually be installed under the direction of the carpentry foreman; however, your crew may be instructed to assist, so you should have some general knowledge of the subject.

The plans will have a "door schedule," which will give the kind, quality, size, and location of each door and the direction it is to slide or swing (see Figure 5.15). A door should operate easily and smoothly and fit snugly but not tightly. It should fit evenly against the doorstop. A warped door should be not installed until it has been examined and approved by the engineer.

The butts (hinges) must be installed in a straight line and set so that the head of the bolt is up; otherwise, the bolt may fall out.

Overhead doors should run smoothly in the tracks, and fit tightly against the stop when in the down position. The bottom should fit snugly against the floor. The balance spring should be set so that the door operates easily.

It is recommended that all swinging and sliding doors be removed once they have been properly installed. The top and bottom edges should be painted, and the door should be immediately rehung. Painters seldom paint the top and bottom edges, and if left unpainted, moisture can enter and cause warping.

5.14. Downspouts

Whenever there is an eave trough or gutter along the edge of a roof, there will usually be downspouts to carry the rainwater to the ground or into a drainpipe. The plan will show the type and how it is to be attached to the structure.

DOOR SCHEDULE

AL – ALUMINUM NA – NATURAL FINISH HM – HOLLOW METAL SCWD – SOLID CORE WOOD

DOOR NO.	DOORS				FRAMES			SADDLE	HDWR SET NO.	REMARKS
	TYPE	SIZE	MAT'L	TRANSOM	TYPE	MAT'L	DETAIL			
1		3^0 x 7^0	AL			AL	AL	D AL		
2		3^0 x 7^0	AL	3^0 X 1^8	D	AL	16A/A5	D AL		BI-FOLD – 1-7/8 THK – ROOT BOT. RAIL EDGE FOR BI FOLD TRACK
3	2	2-2^8 X 7^2	SCWD NF	3^0 X 1^8	A	HM	C	—		
4	1	3^0 x 7^2	SCWD NF	3^0 X 1^8	A	HM	A	—		
5	1	3^0 x 7^2	SCWD NF	3^0 X 1^8	A	HM	A	—		
6	1	3^0 X 7^2	HM	3^0 X 1^8	A	HM	D B	A – VINYL		
7	1	3^0 X 7^2	HM	3^0 X 1^8	A	HM	B	B – MARBLE		
8	2	2-2^6 X 7^2	HM	5^0 X 1^8	D	HM	B D	—		MM FILLER JAMS D-1 ON ONE SIDE ONLY
9	2	2-2^6 X 7^2	HM	5^0 X 1^8	D	HM	B	—		
10	1	3^0 X 7^2	HM	3^0 X 1^8	A	HM	A	D AL		INSULATED HM DRS TRANSOM
11	3	2-2^6 X 7^2 2-3^0 X 7^0	HM	—	C	HM	A	D AL		1, 2, 3, 4 INSUL–HM–DR & TRAN
12	1	3^0 X 7^2	HM	3^0 X 1^8	A	HM	A	D AL		INSULATED HM DR & TRAN
13	2	2-2^6 X 7^2	HM	—	E	HM	B	C VINYL		B LABEL 1-1/2 HR
14	2	2-2^6 X 7^2	HM	—	E	HM	B	A VINYL		SOUND INSUL HM DR
15	1	3^0 X 7^2	HM	3^0 X 1^8	A	HM	A D1	D AL		HM FILLER JAM B INSULATER HM DB ST RAN
16	6	2-2^6 X 7^2 2-3^0 X 1^0	HM	—	D	HM	B	—		1, 2, 3, 4 SEC 13/A4
17	1	3^0 X 7^2	HM	—	B	HM	B	C VINYL		'B' LABEL 1-1/2 HR
18	5	2-1^6 X 7^2	SCWD NF	—	B	HM	C	—		B-1 FOLD DR 1-3/8 THK
19	2	2-2^6 X 7^2	SCWD NF	—	E	HM	C	—		B-1 FOLD DR 1-3/8 THK
20	1	3^0 X 7^2	HM	3^0 X 1^8	A	HM	B	A		B-1 FOLD DR 1-3/8 THK
21	1A	3^0 X 7^2	HM	—	B	HM	A			
22	1A	3^0 X 7^2	HM	—	B	HM	A	21/PS8		

NOTES: 1. Dutch type drs. 2. 2 pairs of doors. 3. Head frame split in two sections for hoist to pass thru. 4. Door frame set to underside of hoist beam.

FIGURE 5.15. Door schedule.

123

5.15. Drainage

Drainage for the site of a completed project will be indicated on the plans and in the specifications. Drainage of a site during construction is the contractor's responsibility and your responsibility in the vicinity of your work.

If you are pumping water from an excavation, you must be sure that the discharged flow is not causing problems on adjacent property and that silt is not filling up catch basins.

If the water being pumped from the site carries much silt, it should be ponded to allow the silt to settle.

There are laws forbidding the discharge of silt-bearing water into streams harboring fish.

If the area requires a permanent system of drainage, the plans and specifications will indicate the system to be used. This can include surface drainage by ditches and swales, or catch basins and piping. There may be French drains with or without piping. (See Fig. 7.5.) Many buildings and other structures may have drainage systems around the foundation footings, as do many roadway pavements. All of these will require that the water be piped to a point of disposal. Many projects will have grade stakes set to indicate the slope, and if so, they must be carefully followed.

The French drains will have some kind of a seal on top of the stone to keep the silt from filling the voids. This seal should be the full width of the trench and should be placed as soon as possible after the stone, with enough backfill to keep the seal in place while the remainder of the backfill is being placed.

When French drains are being constructed, great care must be taken to see that no silt or sand gets into the stone, as enough of the voids could be filled to plug the drain.

5.16. Drawings

The drawings, also called the plans or blueprints, show where and how a structure is to be constructed. You must be sure you have a copy of the latest approved set. Sometimes changes or corrections are made, and revised sheets are issued. If you receive such sheets, be sure you immediately cross out the old design and use only the new one.

When there is a discrepancy between the drawings and the specifications, the specifications govern.

If you are working from a set of drawings, you should retain the set until the work has been approved in order to show your authority for constructing as you did.

5.17. Dredging

Dredging is earth excavation underwater to deepen and/or widen a river or canal or to remove sand and gravel from an underwater pit. The operation causes the mud and silt to become suspended in the water, which kills or drives away the fish. Most states have strict laws governing the disturbing of fish, and some require permits.

Referring to Figure 5.16, the pumper dredge has a cutting device at the bottom of the suction pipe which breaks up the soil and mixes it with the water so it can be pumped through the pipeline. The pipeline will extend to the shore, where the mixture will be deposited within a dried area where the soil will settle and the water will flow back into the river. The pipe will be supported on pontoons. This method is used to make landfill in low and swampy areas and to extend the shoreline out into the water.

The hopper dredge will have suction pipes similar to the pumper on each side, but the discharge will be pumped into hoppers in the dredge. When the hoppers are full, the suction lines are raised, and the dredge will move to an area of deep water where the bottoms of the hoppers are opened and the material falls out. When fully discharged, the hopper doors will close, and the dredge will return to the excavating site.

The dipper dredge is similar to a power shovel in that it has a boom with a dipper at the lower end of the boom or dipper stick. The excavated material is scooped up and deposited on a barge or at another location. The bottom of the bucket has hinges at the back and a latch in front. When the bucket is in position for discharge, the operator pulls on a rope attached to the latch, allowing the door to open. After discharge and the bucket is lowered, the door will close, and the latch will lock it in place.

Note that the pumper and hopper dredges may have stabilizing jacks that can be lowered to the bottom of the river to steady the dredge during operations. The discharge line on the pumper dredge may be assembled and set on pontoons onshore and pulled into the water and attached to the pump on the dredge. Such an assembly will usually be set on greased planks for easy moving.

Dredging can be done by dragline or clamshell, but the excavation will not be as neat as the buckets cannot be seen under water. See Figure 6.24, showing how a dragline operates. Dredging will be the same, except that the bucket will be underwater when loading. The clamshell bucket opens and closes like the shell of a clam, hence the name. The bucket operates from the cable on a crane boom. One cable is attached, so that as the bucket is lowered it opens. The bucket is lowered into the material to be excavated and allowed to settle into it. The first cable is then released, and the second cable is tightened, which closes the bucket and raises it. When the bucket is raised to the required height and swung over the point

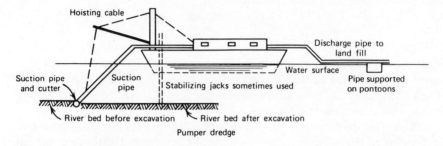

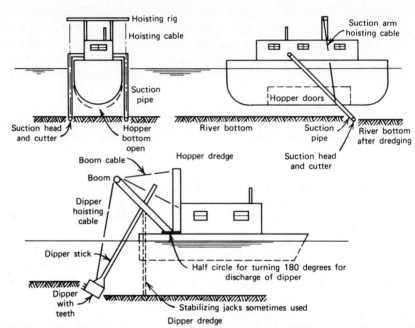

FIGURE 5.16. Dredging. A pumper dredge has a suction line extending to the riverbed, with a cutter device that breaks up the earth and soft rock. The discharge pipe carries the excavated material, suspended in the water, to the point of deposit along the shore. Vertical jacks extend down to the riverbed to stabilize the dredge during operations. The jacks are raised to allow the dredge to be moved. A hopper dredge picks up the material in much the same way that the pumper dredge does, but deposits it in the hold of a ship. When the hold tanks are full, the suction pipe is raised and the dredge moves to an area of deposit. There the hopper doors open, and the excavated material is deposited back on the river bottom at a selected point. A dipper dredge has a dipper similar to that of a steam shovel. The dipper picks up the excavated material from the river bottom and deposits it on a barge for delivery to the disposal area.

of discharge, the operator switches back to the first cable, which opens the bucket. After discharge, the cycle is repeated.

5.18. Driftpins

A driftpin is a hard-steel rod about 8 in. long, tapered at one end to a blunt point. The diameter varies, depending on the diameter of the holes in the steel. It is used to align the bolt or rivet holes in a structural-steel connection. The pin will be driven into one set of holes in a connection to force the plates into position so the holes all line up. Rivets or bolts will be placed in most of the other holes of the connection before the pin is driven out.

5.19. Drilling

Drilling and boring are really the same thing. (See Section 3.25.) Usually, boring refers to making large-diameter holes, such as for tunnels or for sewer or water lines, while drilling refers to smaller holes, such as in drilling for oil or drilling holes for blasting. Drilling also refers to making holes in wood or metal for bolts or rivets.

Drilling steel, concrete, or other hard substances will require high rpm and will usually require water or compressed air to cool the drill. Other materials will require slower rpm.

Efficient drilling requires an experienced operator and sharp tools.

5.20. Drills

Drills (augers) are the cutting units used for drilling or boring. They run in size from ⅛ to 2 or 3 in. in diameter for drilling steel, concrete, and wood and up to several feet for drilling in the earth.

Units for timber are called "wood drills." They will have a short, threaded point for starting the hole, and the cutting edge will be half the diameter of the drill. Those for steel or other hard meterials will have a sharp point with cutting edges on both sides.

Sharpening drills is no easy task and is usually assigned to a worker who is specially trained.

5.21. Drinking Water

Pure, cool drinking water should be available at the work site. Many contractors furnish metal or plastic dispensers which hold one to several gallons of water. They are insulated and keep the water cool for several hours, particularly if kept in the shade. Paper cups should be furnished for sanitary reasons. The expense of furnishing drinking water at the site for each crew is more than offset by keeping the crew content and saving a walk to a centrally located point. The foreman should see that the container is kept clean and full of fresh water.

Workers may experience stomach distress from drinking too much cold water on hot, humid days. One way to prevent this is to put ¼ cup uncooked oatmeal in 5 gal of water, stir it up, and let it sit for about a half hour before using. Some workers do not like the taste, but prefer to drink the mixture and not suffer the discomfort. Try it on a day when the men are hot and tired and see if they like it.

5.22. Drywall Construction

Drywall construction has replaced the old wood lath and plaster method of providing a wall surface. The material used is called sheetrock, which is a plasterboard formed of fibered plaster covered with a moisture-resistant paper. The sheets come in thicknesses of ⅜, ½, ⅝, and ¾ in. There are two types of sheets, the smaller being 48 in. wide and 16 in. long. These sheets are placed with the long side horizontal on the wall studs with the vertical joints staggered. The larger sheets are 48 in. wide with lengths of 8, 10, 12, and 14 ft, so a sheet will reach the full height of a room.

One or two coats of plaster are usually placed over the smaller sheets, while the larger ones will be painted or have wallpaper applied. The larger sheets will have tape applied to the joints. Plaster of Paris or other similar material will be troweled over the tape and allowed to harden, after which it will be sanded, making the joint almost invisible.

Both types of sheets are attached to the wall studs by large-headed galvanized nails. The nails are driven so that the head will be just below the surface, but not deep enough to make a depression. With wood studs, the nails are driven close to the center of the stud. With metal studs, there will be a groove at the center of the stud into which the nails are driven. The sheets can be cut to fit around openings or other objects by cutting the paper with a sharp knife and then bending the sheet, which will break along the line of the cut.

Most drywall work will be done by subcontractors who specialize in this work.

5.23. Ductwork

Ductwork will usually be installed under the sheet metal foreman. The ducts will be either galvanized tin or plastic. They will carry the heated air in winter and the cooled air in summer. If there is more than a few feet in each run, there will be noise-dampening devices in the line. The design and location of the devices will be on the plans. On some projects, the ducts are used to ventilate by bringing in fresh air from outside. (See Section 13.23.) On small jobs, the ducts will be placed within the walls and partitions between the studs or in the ceiling between the joists. On larger projects, the ducts will be suspended from the floor or ceiling joist, and there may be a false or suspended ceiling below. Occasionally, it will be necessary to cut into a structural member to place the duct where required; if so, the plans should detail the method. In the absence of any details, no structural member should be cut without the written permission of the engineer.

The ducts will come to the job site in various lengths, and the joints will be taped to prevent leaks.

5.24. Dummy Joints

When there is a large expanse of uninterrupted wall surface, dummy joints are used for architectural effect. If the wall is of masonry or concrete, some of the joints may also be contraction joints. (See Section 4.39.)

If such joints are called for on your project, study the plans for the method of making the joint and the measurements for layout on the wall so that the desired effect will be accomplished.

5.25. Dur-O-Wal

Most specifications covering masonry walls will require reinforcing for at least the top three horizontal joints. The reinforcing will be heavy-gauge steel wire, a trade name of which is Dur-O-Wal. The smaller gauge will come in rolls, with the heavier gauge in lengths of about 10 ft. There will be special units for the corners. The reinforcing will be about 1½ in. narrower than the width of the wall. The reinforcing should be set in a full bed of mortar with the ends overlapping at least 6 in. This reinforcing will be used mostly in concrete-block walls, but is also used in brick walls.

5.26. Dust Control

Almost all civil engineering construction has some problems with dust. Many specifications require that dust be controlled within reason. Large earth-moving projects usually have a sprinkler truck to keep the dust down, and some use calcium chloride spread over the area.

When wet earth is being excavated and hauled away by trucks on paved streets, water and silt will leak out, unless the truck body is leakproof.

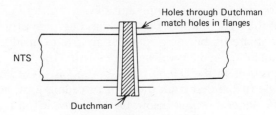

FIGURE 5.17. Dutchman.

The silt will dry, and the traffic will cause a cloud of dust in the area. This is one of the biggest complaints during earth hauling.

There is not much the foreman can do about it if the contractor uses leaky trucks. Usually, the inspector or resident engineer will complain if the problem gets out of hand.

If the trucks have leakproof bodies, the foreman can see that they are not loaded to a height that will allow the earth to spill over the top.

Dust on the floor of a building under construction can be controlled by spreading sawdust, that has a small amount of light oil mixed with it, over the area and then immediately sweeping it up. An exception would be when a special floor finish might not stick because of the oil. In such a case, dampen the sawdust slightly with water.

Dust in tunnels, particularly after blasting, is controlled by forced ventilation. Most state laws will require this.

Dust presents quite a problem while drilling in rock without water flowing to the drill. Workers on such a rig should have some type of respirator. In fact, you and all your crew should have respirators if you are working in dust. If you do not have them, suggest to your superior that they be furnished.

Lung problems can develop several years after a worker has been exposed to dust for even a reasonably short time. Be concerned if your crew is working in a dust-laden area.

5.27. Dutchmen

Figure 5.17 will show you what a dutchman is. It can be of steel or cast or wrought iron. The sides will be ground smooth to match the flanges of the pipe. There will be a hole in the center to match the inside diameter of the pipe and holes around the rim to match the holes in the pipe flanges. When a line of pipe gets slightly off alignment, not enough to allow the use of a standard bend, the flanges will not be parallel, so the joint cannot be bolted up tight. That is when a dutchman is used. Many times it will have to be ground to the correct angle so each side will be parallel with its flange. Once this is done, gaskets can be inserted and the bolts can be drawn up tight.

5.28. Dynamiters (See Section 17.34.)

5.29. Dynamiting (See Section 3.19.)

6

6.1. Eaves

The eaves of a roof are the portions that extend beyond the outside of the wall as an overhang. Some are just plain boards, while others will be decorated for architectural effects. (See Fig. 24.23.)

6.2. Eave Troughs

Eave troughs collect the rainwater as it runs off the edge of the roof and convey it to a point where it will enter a downspout, or the trough may have an open end which allows the rainwater to flow out. The trough may be of metal, plastic, or wood. (See Section 5.14.)

6.3. Effluent

Effluent is the discharge from a treatment or processing plant or from a tank within such a plant (see Fig. 6.1). It is the opposite of influent, which is the flow into the plant or tank.

6.4. Electrical Work

The electrical work will be under the supervision of an electrical foreman who is skilled in his particular trade. The construction foreman will have

FIGURE 6.1. A 24-in. effluent line from a treatment plant that extends 100 ft out on the river bottom is being covered with crushed stone to keep the current from moving it. Note the pipe in the lower right-hand corner. The stone cover will extend about 20 ft on each side of the pipe and cover it to a depth of about 3 ft. The crane picks up the stone with a clamshell bucket and lowers it over the pipe. A diver checks periodically to see whether the stone is. being properly placed.

little to do with the electrical work, except possibly to cut openings in walls and partitions or provide trenches for underground cables. If there is overhead cable or wire work, you may be involved in setting poles or towers.

Some of your power tools, such as saws, drills, etc., may be electrically driven. If so, the power cables should be well grounded and protected from traffic. The safety guards on all equipment should be in place and operating properly. When power tools are set aside temporarily, they should be placed where persons unskilled in their use, particularly children, will not be tempted to use them. All power tools are safer to use when they are sharp and properly lubricated. When not in use, all tools should be stored in their containers. (See Section 16.31.)

6.5. Elevations

(See Section 13.12.)

Correct elevations of the various parts of a structure, such as the top of a footing, the top of a finished floor, or the top of a column baseplate are vital to ensure that the various parts of the structure will fit together.

If you are involved in transferring elevations from control stakes to some part of a structure, check your arithmetic carefully and keep your calculations. The process is simple addition and subtraction, with the use of a surveyor's level and rod if the points are several feet apart and with a carpenter's level if the distance is only a few feet. Referring to Figure 13.6, starting at the bench mark (BM) or control stake which has an elevation of 428.16 ft, you add 5.24 ft to obtain the elevation 433.40 ft, which is the elevation of the surveyor's or carpenter's level, whichever you are using. If the elevation on the structure is 430.20 ft, you will have to measure down 3.20 ft from elevation 433.40 ft. Please note that the figures are in feet and tenths of feet.

6.6. Embankments

An embankment is the piling up of earth to raise the area above the existing ground. It may be for a roadway, railroad, levee, dam, fill around a structure,

FIGURE 6.2. A water truck spraying a highway embankment to produce the optimum moisture content for compacting the soil. Note that the water spraying from the back of the truck reaches the full width of the roadway. The dark material in front of the truck is fresh earth that has just been spread. The truck sprays up and down the area as the earth is spread until the moisture content meets the requirements.

and so on. Usually the specifications will call for a certain amount of compaction, with 95% being the minimum allowed on most jobs. In most cases, the limits of the toe of the embankment will be indicated by slope stakes set by the surveyors.

On large projects, the degree of compaction will be checked by laboratory technicians who will come to the project with their instruments, make their tests, and furnish a written report. You will hear them discuss the "optimum moisture" content, which means the correct amount of moisture in the soil to allow maximum compaction. During dry weather, when there is not enough moisture in the soil to allow the required density or compaction, water is added by sprinkling. (See Section 4.32 and Figure 6.2.)

Almost all embankments will have a slope of "one on one," "one and a half on one," or "two on one," which means one horizontal foot for each vertical foot, one and a half horizontal feet for each vertical foot, and two horizontal feet for each vertical foot. These will be written on the plans as "1 on 1," "1½ on 1," and "2 on 1," (see Figure 20.29).

The specifications will usually limit the classification of soil that may be used in the embankment, and you should see that the material comes within the set limits. The size of stones if often limited to 6 in. diameter as it is difficult to get uniform compaction around stones of larger diameter. Even well graded sand and gravel need some clay or silt as binder. Crusher-run stone or rock (just as it comes from the crusher) will have enough gradation and fines to compact well. Soil with too high a clay content will be difficult to compact uniformly.

Various types of equipment, such as dump trucks, scrapers, and bottom dump vehicles, are used to place an embankment. The earth should be spread in layers, preferably about 8 in., and quickly compacted by sheepsfoot and rubber-tired rollers. This procedure provides a hard surface for the vehicles to keep them from bogging down. When an embankment reaches a height of a few feet, care must be taken to keep the vehicles from sliding over the side. Rather than have the vehicles dump close to the edge, it is better to dump on solid surface and push the soil into position by bulldozer.

The following method is one way of placing an embankment:

1. Remove all sod and topsoil from the area between the slope stakes. Store topsoil as directed.
2. Place embankment material in 8-, 10-, or 12-in. layers as directed. Compact the layer to density required. (See Section 19.17) (See Fig. 6.3.)
3. Place the next layer as shown on Figure 6.3 and compact.
4. Continue with as many layers as required, compacting as before.
5. When the embankment has been completed to the required height, trim the slope by backblading down the sides. To complete the job, see Sections 14.21 and 21.21.

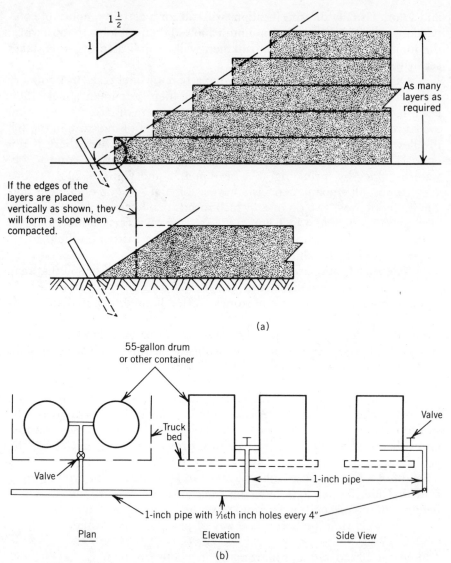

$1\frac{1}{2}$

1

As many
layers as
required

If the edges of the
layers are placed
vertically as shown, they
will form a slope when
compacted.

(a)

55-gallon drum
or other container

Truck
bed

Valve

Valve

1-inch pipe

1-inch pipe with ⅟₁₆th inch holes every 4″

Plan   Elevation   Side View

(b)

FIGURE 6.3. A method of placing embankment in layers. See Section 6.6

6.7. Employees

Usually, the foreman has no control over the selection of the employees
who are assigned to him. It is assumed that the contractor is interested in
the best production possible and so will endeavor to assign the best men
possible to the various operations. It will be your job to learn the capabilities

of each person and place each where they will function the best. (See Section 21.19. Timekeeping.)

Sometimes you will get a misfit who does not fit in with the rest of the crew. He may be lazy or slow to learn and may not keep up his part of the work. She may just not get along with other persons. In any case, if a member of the crew does not fit and holds up the progress of the work, you should not hesitate to confer with your superior, explain the situation, and try to get a replacement. However, you must not criticize a worker for purely personal reasons. (See Chapter 1, Section 5.10.)

6.8. Epoxy (See Section 17.1.)

6.9. Equipment (Contractor's)

Proper equipment, in good working order, with skilled operators, is one of the secrets of successful contracting. The foreman's concern is with the efficient use of each piece of equipment in his care. Usually, the contractor will hire operators who are skilled and know the capabilities and limitations of the unit. The foreman must see that conditions on the site are such that each piece of equipment can operate as near to full time as possible.

Usually, one piece of equipment works in conjunction with one or more other units, so the breakdown or delay of one unit can affect the production of several others. A backhoe will have several trucks and one or more dozers; scrapers will have a dozer to help load; concrete pumps will have several Redi-mix trucks to supply them with concrete; bituminous-concrete spreaders will have several trucks and rollers; placing an earth embankment will require trucks and/or scrapers, graders, dozers, various types of rollers, and sprinkler trucks. Carpenters on civil engineering projects seldom use hand-powered tools. The circular saws, table saws, drills, planes, nailers, chisels, and jackhammers will all be power driven. If electric, there will be cables to each unit which must not interfere with other operations. Saw blades, drill bits, and chisels must all be sharp at the start of work, and extra ones should be available if needed. Gasoline-engine-powered tools will need fuel, oil, and lubrication. (See Sections 7.16. and 14.1.)

See Figures 6.4–6.21.

Backhoes

Backhoes (pullshovels) come in several sizes from $\frac{1}{4}$ to 5 yd,^{3} but the most common on civil engineering work are the $\frac{1}{2}$–$\frac{3}{4}$ and 1 yd^3 sizes. The $\frac{1}{2}$ yd^3 can dig a trench about 6 or 7 ft deep, but is not heavy enough to remove hard clay, hardpan, or shale without pounding with the teeth on the bucket. This is not good for the machine, as it increases the wear and tear and can cause breakage. The $\frac{3}{4}$ yd^3 unit can dig a trench up to 10 ft deep if the

FIGURE 6.4. A typical construction crane used on small projects.

soil is easy to loosen. When the boom of a backhoe is extended down the full length, it loses much of the leverage needed to loosen the soil. The ¾ yd³ unit is a powerful machine and can loosen rather hard material. The 1 yd³ unit, due to the weight of the boom and bucket, can loosen almost anything but hard rock. Regardless of the ability of the backhoe, it is best not to use it to its maximum strength, but rather to loosen the material with a jackhammer or by preshooting. A backhoe is used effectively when it can make a pass, get a full bucket, and swing quickly to discharge.

Scrapers

Scrapers (pans) should load quickly, make a short run to the point of deposit, and return. They can do this only in soft soil. If the loading gets tough, a

FIGURE 6.5. Front-end loader on caterpillar track.

FIGURE 6.6. A heavy-duty front-end loader by Trojan. This model is hinged between the axles so that the whole front section turns as a unit. The unit picks up earth, sand, gravel, and so on, and deposits it in a truck, places it in a stockpile, or spreads it on the ground.

FIGURE 6.7. The particular model of front-end loader in Fig. 6.7 has an added feature that allows the bucket to tilt sideways to deposit lining in the bucket of a backhoe.

bulldozer pushing at the rear can speed operations. Harder soil may require the use of a scarifier or ripper. The scarifier is a metal frame holding a row of hardened-steel teeth. The unit will be attached to a road grader or the back of a power roller. The ripper is similar to a heavy farm plow, except that instead of a plow blade it will have a hardened-steel tooth 2–3 in. wide and 4–6 in. deep. It may be straight or curved. The ripper will usually be on a wheeled frame and pulled by a bulldozer. The scarifier will break up old subgrades, unpaved roads, and other such hard surfaces. The ripper will break up bituminous paving and shallow rock.

Mixers

Most concrete is delivered in Redi-mix trucks from a commercial supplier, unless the project is too far from the supplier's plant. In that case, the contractor will have his own mixing plant, the size of which will depend on the amount of concrete needed daily. On a project requiring several hundred cubic yards of concrete per day, there will be an automated plant with computerized controls that measure the correct amount of aggregate, cement, and water. When lesser amounts are needed per day, a 1 yd^3 or less mixer may be sufficient. These smaller mixers will usually be hand loaded. A standard 1–2–4 mix will be used, and the sand and stone will be fed into the mixer hopper by wheelbarrows. When exact quantities are required, the wheelbarrows will be partitioned off so as to take 2 ft^3 of stone and sand. This is done by placing a 2 × 8 in. plank section in the back of the barrow so it will hold exactly 2 ft^3 by strike-off measurement. Then two barrows of stone, one barrow of sand, and one sack of cement will provide the correct mix. The quantity can be doubled or tripled as required by using four barrows of stone, two of sand, and two sacks of cement, and so on, depending on the size of the mixer.

Any mixer should be thoroughly cleaned immediately after use. This is easily done by placing about 1 ft^3 of crushed stone and plenty of water in the mixer and letting it rotate for several minutes. Crushed stone works better than gravel because of the sharp edges.

Bituminous-concrete spreaders should be cleaned immediately after use by spraying the affected areas with kerosene and scraping the surface with a square-point shovel or wide-blade putty knife. Cleaning the screw conveyor in the hopper is the most difficult.

FIGURE 6.8. A backhoe, with a ¾ yd bucket, excavating a trench for a storm sewer. This is the machine most commonly used for trenching for sewer and water lines. There are several different makes on the market, but there is not much difference between them. This machine can dig effectively to a depth of about 10 ft. It can excavate reasonably hard material, including soft shale, but for hard shale, hardpan, or rock it will require preshooting. (See Section 17.38. Preshooting Rock.)

FIGURE 6.9. One of the most versatile pieces of equipment found on a construction site. One end is a small backhoe, and the other end is a front-end loader, which can be used as a bulldozer. It is very maneuverable, can work in close quarters, and moves quickly from site to site. Several manufacturers make them, but they are all basically the same.

FIGURE 6.10. A bulldozer from the Caterpillar Company.

Cherry Pickers

Cherry pickers used to be small, easily maneuvered units good in tight places, but now they will rival most cranes for boom height and reach. They are still distinguishable by the solid, retractable boom. The larger ones will have stabilizing outriggers the same as cranes and will be used in very much the same way. They will require the same maintenance as cranes for both cable and hydraulic units.

Power Tools

Jackhammers are cleaned inside by removing the air hose and pouring in about ½ cup of kerosene, reconnecting the hose, and operating the hammer a few seconds to blow the kerosene out through the exhaust ports. The openings of the exhaust ports should be kept clean. There should always be an oiler on the air line to keep the hammer lubricated when in use.

Electric tools such as drills, saws, and planes should have a grounding-wire connection. They should never be left unattended, and the cable should be laid out so as to be away from areas where equipment will be moving about. Sharp tools are much safer than dull ones.

Construction equipment left out-of-doors during extremely cold weather will be hard to start in the morning. Operators used to place a couple of

FIGURE 6.11. A bulldozer from the John Deere Company, shown backfilling a trench.

FIGURE 6.12. A "sheepsfoot" roller. This one requires a tractor to pull it. Note that there are two sections for ease in turning a unit this wide. Weight can be added by filling the rollers with water.

The sheepsfoot roller is used on soil with a high clay content. At the start, the roller is used without added water, and the "feet" will penetrate the soil to their full depth. As the soil is compacted, the feet penetrate less and less until they sink in less than an inch. Water is usually added when the feet penetrate about 2 in. and at that time a rubber-tired roller is used along with the sheepsfoot to provide a smoother surface.

drops of ether in the engine intake, but today you can purchase a can of starter fluid to spray into the intake. You should follow the manufacturer's instructions.

Cranes (Mobile)

Cranes will have charts listing the maximum loads they can lift at various angles of the boom. Some booms will have an angle indicator, and some will have outriggers to stabilize the unit when handling heavy loads and when swinging to either side.

All cables on a cable unit should be checked at regular intervals for broken strands or other damage, and the cable should be replaced if necessary. Hydraulic units should have their pipes and hoses checked, particularly at connections, and any leaks should be corrected. The brake lining on the drums of all cable units should be checked and replaced as necessary.

In built-up areas where the rear of the crane housing will swing into a traffic lane, a flagman must be provided to direct and warn the traffic.

In areas where overhead power cables are present, care must be taken to keep the crane boom at least 10 ft from the cables. Should the boom touch a cable and the operator not be affected as he sits on the seat, warn him not to move, but to sit still with arms folded and not touch anything. Call the power company at once and give them your exact location. Keep all other workers away from the machine. It is a good policy to confer with the local power company, informing them where you will be working and asking for any special suggestions they may have on preventing accidents.

Crane units may move on caterpillar tracks or be on heavy-duty truck chassis.

Gradall

The gradall is similar to the backhoe, except that the boom is solid, with telescoping sections like those on the cherry picker. The bucket at the end of the boom can swivel about 180 deg. Such a machine is very useful for shaping ditches and slopes behind the ditches. Many highway departments use this machine with a wide bucket to clean out ditches.

FIGURE 6.13. A caterpillar pan or scraper. This one has two power units with a hinge between them. Such a unit has more power for loading. The unit is loaded by moving forward while sitting on the ground, as shown. When the pan is loaded, it is raised, and the unit is moved to the point of discharge. To discharge, a wall in the back of the pan moves forward, forcing the soil out. The depth at which the soil is deposited is governed by the speed of the unit and the moving wall.

FIGURE 6.14. A gang-type vibrating compactor. Six vibrating units are side by side and operate as a single unit. The motor operates a generator, which supplies electricity for the vibrators and power for the caterpillar treads that propel the unit. The chains that run from the vibrators to the pulleys on the rod are for elevating the units when moving from one area to another. This unit is compacting an 8 in. layer of No. 3 stone for the base course of an airport taxiway.

FIGURE 6.15. A road grader with the blade down and at an angle. This is the position the blade will be in for most roadway surface forming.

FIGURE 6.16. Another grader, this one with a cab and scarifier teeth in front of the blade. These teeth are lowered to break up hard surface or old paving.

FIGURE 6.17. A heavy-duty dump truck used on large earth-moving projects. The short wheel base makes them easy to maneuver.

Pumps

On civil engineering projects, both diaphragm and centrifugal pumps are used, and occasionally the piston-type is used for pumping concrete or other heavy substances.

A diaphragm pump will have a rubber or composition diaphragm, which causes suction as it rises and discharges as it is lowered. There will be a rubber ball valve at both the suction and discharge ends. The seats

FIGURE 6.18. Front-end loader loading a dump truck.

for these ball valves must be cleaned occasionally to keep them from leaking. Such pumps will pass small stones, sticks, and other debris, along with heavy concentrations of mud in the water. The suction hose will be reinforced with steel bands to keep it from collapsing. There will be a valve at the intake end to keep the water from draining out during the discharge stroke. The discharge hose may or may not be reinforced. Diaphragm pumps are self-priming and will hold the prime a long time if there are no leaks in the diaphragm.

The centrifugal pump will have a fan-shaped impeller with the blades fitting close within the housing so that very small hard objects can pass through. The suction end should have a screen to keep out sticks, stones, and so on. Some such pumps are self-priming once they have been primed at the start of the day. Others must be primed after each stop. There will be a plug or valve on the top of the housing through which the prime water is poured. There should be enough water to fill the pump housing half full.

Piston or plunger pumps are used to pump concrete and grout. These pumps are an excellent way of moving concrete, as there is no spilling along the way and they deliver the concrete to the point where it is needed. However, there can be problems if the weather is very hot or very cold and the concrete is pumped a long distance. In hot weather, the concrete will set up if left standing in the pipe or hose for very long, and in winter weather it can freeze. In either case, it is better to waste some concrete by pumping out the line, flushing it with water, and starting over when ready.

Pile Drivers

All pile drivers will have "leads," which are open frameworks of timber with steel guides for the hammer. Some leads are all steel. The leads will be

40–50 ft high or more, long enough to receive the longest pile expected. The leads may be "swinging," which means they will be hanging from a cable of a crane, or they may be "fixed" leads built into the framework of a unit holding the leads, the equipment for hoisting the hammer and the pile, and the power unit. The power unit may be a steam boiler or a gasoline or diesel motor with an air compressor. Most will be on caterpillar treads, although some of the old steam units will still move over rollers.

All hammers should have a "follower block," which is a device for protecting the top of the pile during driving. It will be a cast-iron unit with a round slot to receive a 6-in. block of wood, which acts as a cushion as the hammer strikes. The wood blocks are replaced as they wear out. The follower block will sit on top of the pile and follow it down, hence the name. (See Sections 9.1. 17.15.)

Graders

Road graders are manufactured by several companies, but all are similar. Most will have two rubber-tired wheels in front and two rubber-tired wheels in tandem in the rear where the power is applied. Between the wheel axles will be a hard-steel blade 8–10 ft in length and concaved towards the front. The blade will swing in an arc of about 60 deg and can be raised or lowered as needed. Some models will have front wheels that can slope to add pressure when blading hard surfaces. Most of the graders will have provisions for

FIGURE 6.19. An installation of raw sewage pumps. The suction pipes extend down through the grating into the well that receives the sewage after it has passed through the bar screen and grit chamber. The sewage is pumped from here to the primary settling tanks. Note the control panel on the right and the flow recording units on the background wall.

FIGURE 6.20. Lime-solution mixing tanks. The dry lime from the storage room above is conveyed to the tanks, where it is mixed with water to form a slurry. The slurry is pumped to various treatment tanks on the site by means of the diaphragm pumps seen in the foreground.

attaching scarifiers. They may be in front of the blade or at the rear of the machine.

The blade will be at a right angle to the centerline of the grader when spreading material and at an angle if shaving off a subgrade that is too high.

The graders are used to spread roadway base-course materials for rolling, to blade a subgrade to the proper line and grade, to spread the bituminous materials when shimming an old roadway for resurfacing, to spread embankment material as it is deposited by dump trucks, and to remove snow from a roadway if the depth is not too great.

Rollers (Compactors)

Rollers for construction work come in several types. The original "steamroller," which is now propelled by an internal combustion motor, gasoline or diesel, has one wide roller in front and a narrow one on each side at the rear. There is also a tandem roller with two wide rollers and a tandem roller with a third roller which can be raised or lowered to provide extra pressure. These rollers have steel drum wheels that can be filled with water to increase their weight. There is also a roller with one wide steel roller in front and two rubber-tired wheels in back. Such units may be equipped with vibration on the steel roller.

The original and tandem rollers are used to compact bituminous-concrete paving and will have a sprinkler to keep water on the roller so as not to pick up the paving material. When used to compact earth or stone, there will be a steel scraper to scrape off mud or stone chips that stick to the roller.

	DATE	LOCATION			AM'T	Total
		Conte 1B			C Y	
		ITEM 32 Excavation below grade				
1	4/8/77	Ridgeway	0+30-2+13	183'x3x.5	10.2	
2	4/9	"	2+13-4+20	207/3/1.5	11.5	
3	"	"	4+20-5+18	98/3/.6	6.5	
4	4/11	"	5+18-7+36	218/3/1	24.2	
5	"	"	7+36-9+56	220/3/1.5	12.2	
6	4/12	"	9+56-11+77	221/3/1.8	19.6	
7	4/13	"	11+77-12+97	120/3/1	26.6	
8	"	"	0+00-0+77	77/3/2	8.5	
9	4/14	"	0+77-3+50	273/3/1	30.3	
10	4/15	"	3+50-4+46	96/3/1	10.7	
11	4/16	"	4+46-5+85	139/3/.6	9.3	
12	4/18	"	5+98-7+87	189/3/.6	12.6	
13	"	"	7+87-8+54	67/3/1.5	11.2	
14	"	"	8+54-8+95	41/3/2.5	11.4	
15	"	"	8+95-9+98	103/3/2	22.9	
16	4/19	Rte 48	10+20-11+45	125/3/1.5	20.8	
17				Estimate N° 6 –		248.5
18	4/20	"	11+45-12+15	70/3/1.5	11.7	
19	"	Meridan	0+00-2+15	215/3/1.8	19.1	
20	4/21	"	2+15-2+43	28/3/.5	1.6	
21	"	"	2+43-3+81	138/3/.8	12.3	
22	"	"	3+81-4+11	30/3/2	6.7	
23	"	"	4+11-4+24	13/3/1	1.4	
24	4/22	"	4+24-4+76	52/3/2	11.5	
25	"	"	4+76-7+17	241/3/1	26.8	
26	4/23	Rte 48	1+25-2+98	173/3/1	19.2	
27						
28	4/27	Rte 48	4+94-5+33	39/3/3	13.0	
29	"	"	0+00-1+22	122/3/2	27.0	
30	"	"	0+00-0+56	56/3/1.5	9.3	
31	4/28	"	0+56-3+36	180/3/1.5	30.0	
32	4/29	"	3+36-4+24	88/3/1.5	14.7	
33	"	"	4+24-6+32	208/3/1.0	23.1	
34	4/30	"	6+32-8+10	178/3/1	19.8	
35	5/2	"	8+10-8+84	74/3/1	8.2	
36	"	"	8+84-10+08	124/3/.8	11.0	
37	5/3	Bencon G R/w	0+00-0+86	86/4/1	12.7	
38	"	"	0+86-1+50	64/4/2	18.9	
39	"	"	1+50-2+27	77/4/1.5	17.1	
40	"	Fine Hec	1+02-3+07	168/3/1	24.7	
				Estimate N° 7 –		588.3

FIGURE 6.21. Quantity work sheet. For each item of the contract (in this case, Item 32, Excavation Below Subgrade), there is one or more sheets. Each day the amount of work accomplished under each item is entered on the sheet, giving the location and the amount of work. At the end of the month, the work listed is totaled to arrive at the amount to be entered on the monthly estimate.

Regardless of what is being compacted, each pass of the roller should overlap the previous pass by 50%.

There is also a "sheepsfoot" roller (see Fig. 6.2). The "feet" are about 6 in. long, and when starting to compact earth, the feet will sink in for the full depth. As the earth is compacted, the feet will penetrate less and less until they are almost on top. They will always penetrate a small amount. As the compacting progresses, the rubber-tired roller is used with the sheepsfoot to produce a smooth surface. The rubber-tired roller will have several rubber-tired wheels, usually six or eight on each of the two axles. The air pressure in the tires is important to get the proper compaction. (See Fig. 6.13.) Some rubber-tired rollers will have wheels that wobble to increase the compaction. Both the sheepsfoot and rubber-tired rollers will be pulled by a tractor.

6.10. Equipment (Installed as Part of Project)

Most all projects will have some kind of equipment installed. It could be only heating and ventilating equipment, or it might be more sophisticated process equipment. In either case, it must be properly installed to function correctly. The more complicated units will have installation instructions attached, and these should be followed, or they may require a factory-trained mechanic to install and test-run them. Such units will usually have operating instructions and even a supply of grease and small adjusting tools included; if so, these should be carefully stored where they will be available when needed. Most equipment will require anchor bolts, and the foreman must see that these are installed in the proper position and with the correct amount projecting above the base. Once in place, anchor bolts should be protected until the equipment is set.

The base of the equipment should bear evenly on the foundation, and the anchor bolts must be tightened evenly so as not to put an uneven strain on the base. An uneven strain on the equipment base could cause it to warp, which in turn could cause problems with the shaft and bearings.

Newly installed equipment should be turned over by hand first to be sure it does not bind. If it binds, the factory mechanic should be called to check it. If the equipment turns over freely and is electrically powered, the switch should be turned on to see that the equipment rotates in the proper direction. If the rotation is reversed, the electrician should be called to change the wiring to the proper positions.

When new equipment is first put in operation, it should be carefully observed to see that the bearings do not get too hot. They should be very warm, but not hot. They should produce a "hum," not a "chatter."'If the bearings chatter, they should be inspected by a factory-trained mechanic and corrected before further operation. In a noisy room, it is hard to detect a bearing chatter, so it is best to use a stethoscope.

All equipment should arrive on the job site properly lubricated, but sometimes they are not, so each unit should be checked.

6.11. Erosion

Erosion can affect both "cut" and "fill" operations, but will be more extensive on the fill. The degree of erosion will depend on the extent and slope of the exposed earth.

Protection from erosion is provided in several ways, such as by riprap, spreading mulch, placing a blanket of jute or other fiber, sodding, and concrete or bituminous surfacing. See each of these methods under its own heading. Erosion is always a problem when placing an embankment during rainy weather. You should try to channel off the surface water and keep as much as possible from running down the embankment slope.

Erosion in ditches on a steep slope can be retarded by placing small bales of straw across the ditch at intervals to provide a series of small dams to retard the flow. Highway ditches on steeps can be lined with crushed stone and sprayed with hot asphalt.

6.12. Escalators

Escalators are seldom used on civil engineering projects, and if they are, they will be installed by a subcontractor or equipment dealer.

Some general things you should know about escalators are that they should operate smoothly without jerking, there should be no binding where the treads pass through the grating at the top and bottom, the treads should be level as they ascend or descend, and there should be no vibration felt by the feet.

6.13. Estimates

The contractor is paid according to monthly estimates of the work done to date as the job progresses. The foreman is not usually involved in the preparation of the estimate. However, he can help by seeing that his work is kept up to date by fully completing each phase of the work as much as possible. On work involving trenching or other earth work, cleanup is always a problem. Surface restoration should be done as soon as possible. Many engineers hold up all or part of the trench payment until all cleanup is approved. (See Figs. 6.22 and 6.23.)

TOWN OF _____, NEW YORK
WATER POLLUTION CONTROL PLANT
CONTRACT NO. 2A—GENERAL
SEWER DISTRICT

TO: TOWN OF _____

MONTHLY ESTIMATE

Pursuant to the terms of the Contract dated March 31, 1977, by and between John R. _____ Construction Co., Inc., Contractor, and the Town of _____, Owner, for the construction of Contract No. 2A—General, Water Pollution Control Plant, _____ Sewer District, we submit herewith the Third Monthly Estimate for work done to August 23, 1977.

Item No.	Description	Bid Quantity	Quantity To Date	Unit	Unit Price	Bid Amount	Amount To Date
1.	Clearing	Lump sum	100%	L.S.	L.S.	$ 60,000.00	$ 60,000.00
2.	*Excavation*						
2a.	General	2000	8,979	c.y.	$ 15.00	30,000.00	134,685.00
2b.	For structures	8600	6,633	c.y.	15.00	129,000.00	99,495.00
2c.	Below subgrade	300	—	c.y.	15.00	4,500.00	0.00
3.	Yard piping	Lump sum	—	L.S.	L.S.	40,000.00	0.00
4.	Special backfill	1000	—	c.y.	8.00	8,000.00	0.00
5.	Lining	450	—	c.y.	8.00	3,600.00	0.00
6.	Earth embankment	2000	3,886	c.y.	15.00	30,000.00	58,290.00
7.	Concrete (Class "C")	100	—	c.y.	35.00	3,500.00	0.00
8.	Sheeting & bracing left in place	5	—	MBF	400.00	2,000.00	0.00
9.	Reinforcing steel	2000	—	Lb.	0.30	600.00	0.00
10.	Painting & finishing	Lump sum	—	L.S.	L.S.	15,000.00	0.00

No.	Description		Est. Qty	Qty	Unit	Unit Price	Amount
11.	Comminuting equipment	Lump sum		—	L.S.	$ 15,000.00	$ 0.00
12.	Raw sewage pumping station	Lump sum		—	L.S.	80,000.00	0.00
13.	Chlorination equipment	Lump sum		—	L.S.	12,000.00	0.00
14.	Miscellaneous equipment	Lump sum		—	L.S.	10,000.00	0.00
15.	Process instrumentation	Lump sum		—	L.S.	20,000.00	0.00
16.	Grit chamber	Lump sum		—	L.S.	10,000.00	0.00
17.	Control building	Lump sum		—	L.S.	75,000.00	0.00
18.	Interior process piping	Lump sum		—	L.S.	20,000.00	0.00
19.	Interior process valves & gates	Lump sum		—	L.S.	20,000.00	0.00
20.	Combined treatment tanks	Lump sum		—	L.S.	654,758.00	0.00
21.	Chlorine contact tank	Lump sum		6	L.S.	10,000.00	0.00
22.	Sludge drying beds	Lump sum		18	L.S.	50,000	14,000.00
23.	Special structures	Lump sum		—	L.S.	5,000.00	0.00
24.	Pumping equipment	Lump sum		—	L.S.	20,000.00	0.00
25.	Blower equipment	Lump sum		—	L.S.	20,000.00	0.00
26.	Concrete steps & sidewalks	Lump sum		—	L.S.	2,000.00	0.00
27.	Engineer's field office trailer	Lump sum		—	L.S.	2,000.00	0.00
28.	Project sign	Lump sum		100%	L.S.	500.00	500.00
29.	Fencing	Lump sum					
30.	*Roadwork*						
30a.	Unclassified excavation		1000	245	c.y.	$ 15.00	3,675.00
30b.	Subbase course		1500	—	c.y.	8.00	0.00
30c.	Base course		1000	—	Tons	22.00	0.00
30d.	Top course		250	—	Tons	22.00	0.00

FIGURE 6.22. A copy of a monthly estimate. Notice that on this contract Excavation Below Subgrade is Item 2C.

(*Figure continued on p. 136.*)

Item No.	Description	Bid Quantity	Quantity To Date	Unit	Price Unit	Amount Bid	Amount To Date
31.	Top soil	1200	—	c.y.	4.00	$ 4,800.00	$ 0.00
32.	Seeding	7200	—	s.y.	1.00	7,200.00	0.00
33.	Hoisting equipment	Lump sum	—	L.S.	L.S.	10,000.00	0.00
34.	*Miscellaneous metal*						
34a.	Aluminum	500	—	Lb.	3.00	1,500.00	0.00
34b.	Steel & wrought iron	1000	—	Lb.	0.40	400.00	0.00
	Ineligible Items						
35.	Bradley Avenue water main	Lump sum	—	L.S.	L.S.	20,000.00	0.00

TOTAL $1,465,858.00 $370,645.00

Advanced for materials on site per receipted invoices   9,723.01

Total amount of work done to date $380,368.01
Total Amount of Work Done to Date $380,368.01
Less 10% retained   38,036.80

     $342,331.21

Less amount certified—previous estimates   98,068.50

AMOUNT DUE CONTRACTOR—ESTIMATE NO. 3 $244,262.71

We hereby certify that the work listed above has been done, to the best of our knowledge and belief, in accordance with the Contract and Specifications.

FARMERS HOME ADMINISTRATION   Consulting Engineers   , P.C.

By: _____ Date: _____   By: _____ Date: _____

Figure 6.22 (*continued*).

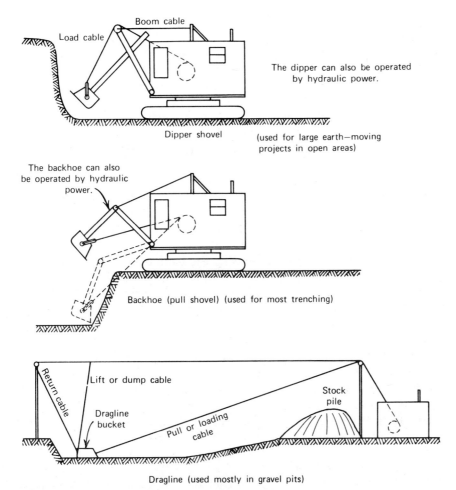

FIGURE 6.23. Types of shovels. The dipper shovel (power shovel) can be powered by cable or by hydraulic power. This type of shovel is used mostly on large earth-moving projects in dry, stable areas. The backhoe (pullshovel) is the most common excavator used today. It ranges in size from ½- to 5-yd³ capacity. It is particularly adapted to trench work. The dragline is most frequently used in gravel pits where the equipment is moved only rarely.

6.14. Excavation

On civil engineering projects, the earth excavation will be done by machines.

The type and size of machine will depend on the kind and size of the excavation. (See Figure 6.23.)

Excavations for buildings will usually be done by backhoe, power shovel, or front-end loader (See Section 6.9.) The excavated material will be loaded onto trucks and hauled off the site or may be placed in low areas on the site.

The limits and depths of the excavation will be controlled by use of batter boards or, on large jobs, survey instruments set over control points. It is the foreman's job to see that the excavation stays within the limits intended.

Truck drivers should be instructed to place their trucks where they can be quickly loaded with a minimum of maneuvering by the excavating equipment. As a loaded truck moves away, the next truck should move up quickly into position. If the trucks will travel over a highway, they should not be loaded to the point where earth will spill off onto the road surface. Such spills can cause traffic accidents by being slippery when wet or causing dust clouds when dry. The spills are also illegal in most states. Gravel on the road surface can be thrown by tires and injure pedestrians and damage windshields and headlights.

If the excavating equipment has enough reach, it is best to keep the trucks outside the excavation. The trucks must stay far enough from the edge of the excavation to avoid causing the bank to collapse.

On large projects, the trucks will have to get onto the area of virgin soil and may have difficulty maneuvering, particularly during and after a rain or in an area of springs. (See Section 20.44.). It may be necessary for a bulldozer to push the truck to solid ground. A ramp up out of an excavation may require crushed stone on the surface for traction.

Excavation of high ground to lower the contours and to grade an area should start at the lower, downhill area. This allows for better drainage, so the surface will stay more stable and the equipment will be less liable to bog down. Most such excavation is done with power scrapers or "pans" when the excavated earth is to be moved less than a mile. For longer distances, large dump trucks or bottom-dump power wagons are used. (See Section 6.9.)

Trench excavation is done by a backhoe or trenching machine (see Section 6.9.) As most trenches are backfilled as soon as the pipe, cable, conduit, or whatever is being installed is in place and inspected, the excavated earth will be piled on the surface along the trench and bulldozed back into the trench. The spoil pile should be back from the edge of the trench far enough to avoid causing collapse of the bank. (See Section 21.28.)

Earth-handling equipment like dump trucks, backhoes, front-end loaders, and bottom-dump vehicles should be thoroughly cleaned at the end of the day to remove earth deposits that build up during use. These deposits reduce the capacity of the unit and add to its weight. After all deposits have been removed, the metal should receive a coat of light oil. Old crankcase oil will work.

When a large excavation becomes flooded from rain storms and there is lower ground in a nearby area, an economical method of dewatering is to use a siphon. There must be a drainage area where the water can be discharged without damaging other property. (See Section 20.26.)

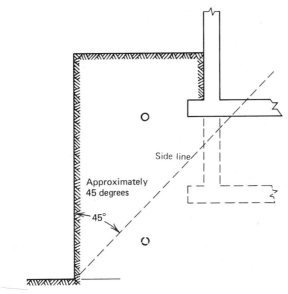

FIGURE 6.24. Slide angle.

When starting an excavation requiring tight sheeting, the following method is simple and fast. Refer to Fig. 20.18. Excavate 2 ft or more, the depth depending on the stability of the soil. Install the wales and struts in the proper alignment and start the sheeting tightly against the wale. Proceed with the excavating, driving the sheeting as the work progresses. (See Figure 6.24.)

6.15. Exfiltration

Exfiltration is a method of testing for leaks in a sewer line where there is no groundwater or the groundwater is too low to cover the pipe. A sewer bag is inserted in the pipe on the downstream side of the lower manhole and in the upstream side of the higher manhole. Water is placed in the line and up into the upstream manhole to a height that will give the head of water required. A mark is placed at the high-water point and observed for a specified time, after which the drop in water level, if any, is noted. If there is a drop in the water level, the volume can be calculated and checked against the allowable leakage. If the leakage is more than allowed, the pipe will have to be checked for leaks and be repaired.

Most specifications will allow a certain amount of leakage.

6.16. Existing Underground Facilities

The plans the contractor uses to prepare his bid may indicate underground facilities such as sewers, water lines, power cables, and foundations of abandoned or removed structures. A note may state that the positions on the plans are only approximate and that the owner assumes no responsibility for their exact position. The contract may say that the contractor will be paid certain amounts for the removal of stated items, or it may say that the contractor must cover the costs under items in the contract.

If you are involved with excavating and run into conditions not clearly indicated on the plans, confer with your superior on whether conditions warrant submitting a claim for extra compensation. If there is doubt as to whether a claim would be approved, it is better to make the claim and have it turned down than to miss a chance to have it approved.

Information on which to base a claim should be complete, showing what was done, how much labor, equipment, and materials were used, and where on the project the work was done.

6.17. Expanded Metal

Expanded metal has very sharp ridges and should be handled with heavy gloves. The light-gauge metal is used as reinforcement for lightweight concrete, stucco, and plaster. The heavier gauges are used for screens guarding machinery and openings in floors, and also for windows, doors, and louvers. As the metal sheet is punched and expanded, it will have sharp ridges on one side. This side should be placed away from traffic. (See Figure 6.25.)

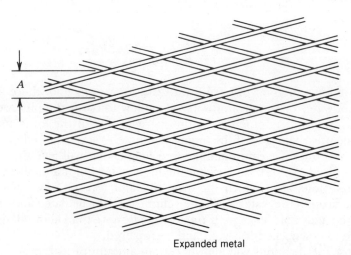

Expanded metal

FIGURE 6.25. A, Opening will depend on the gauge of the sheet from which it was expanded.

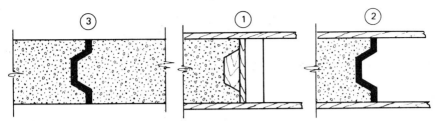

FIGURE 6.26. Expansion joint. (1) The first concrete pour forms half of the joint. (2) The form is then removed, and the premolded expansion material is placed against the concrete. (3) The finished joint.

6.18. Expansion Joints

All materials expand and contract with changes in the temperature. Without expansion joints, the material would crack in irregular lines, which are not only unsightly, but could occur at points of maximum stress in the structure.

If you are involved in placing joint material, study the plans for details. If the joint is in a poured-concrete wall, the form to make the joint must be strong enough to resist the pressure of the wet concrete. It must be set so that the resulting groove in the concrete is straight and smooth, so the premolded joint material will fit tightly against the concrete. If there is a water stop in the joint, it must be set so that the forms can be removed without damaging the stop. The premolded material and the stop must be secured so that when the next concrete is poured it will not displace them. On concrete paving, the rods in the expansion joint must be set parallel to the centerline of the slab so that the concrete will slide easily on them. The expansion joints in masonry must be kept clear of mortar as the brick or block units are installed. Expansion joints across the roadway of a bridge will have steel-plate covers which must lay flat and be parallel to the surface of the roadway; otherwise they will be noisy in traffic and may be bent so as to bind. Expansion and contraction in metal pipes above ground are usually controlled by loops in the pipeline. However, some pipelines have slip joints, which must have some protection to keep the joint clean when contracted; otherwise, it may bind the next time it expands. (See Figure 6.26.)

6.19. Expediting

Expediting the progress of the work is fine if properly done. However, there is a point at which the speedup will result in inferior work and accidents, both of which could cost much more than the savings from expediting.

Of course, any good foreman will plan to carry out his assignment as promptly and correctly as possible. But if you are being urged to speed up to a point where you feel you cannot do your job correctly, you should confer with your superior, be guided by his attitude, and do the best you can.

6.20. Explosion Proof

In sewage treatment plants and other processing plants, there will be areas where explosive gases will, at times, leak into the air. At such locations, explosion-proof electric fixtures will be installed. The fixtures will be heavier than standard ones, and all glass will be extra heavy. There will be rubber gaskets at all joints, and the fixtures must be down tightly against the gaskets. All glass must be checked for cracks before being installed.

6.21. Extra-Work Claims

(See Section 4.20.)

Extra-work claims are presented to the owner by the contractor when he believes he is entitled to extra compensation for various reasons. He may claim that the plans were unclear and that the work is costing more than he anticipated. He may claim that the underground conditions were worse than he could have reasonably expected. If you believe your crew is having to do more than would normally be expected, confer with your superior.

If you believe your work may be the basis for a claim, be sure to keep a careful record of all labor, equipment, and material used.

6.22. Extra-work Orders

Extra-work orders cover extra work required to complete some part of the original plans. Unusual weather or unexpected underground conditions are two reasons for an extra-work order. The owner recognizes that the contractor could not reasonably be expected to have anticipated the problem and agrees to pay the extra costs.

The extra-work order will show what was done and why and will detail how the contractor is to be paid.

Extra-work orders can also cover additional work on items already in the contract, such as on more fencing or roadway as shown in Section 4.20.

EXTRA WORK ORDER NO.: 8   DATE:

CONTRACT NO.: 3

PROJECT: _____SANITARY DISTRICT   FILE: 55.27

OWNER: _____Department of Sanitation.

CONTRACTOR: _____

AUTHORIZATION IS HEREBY GRANTED FOR THE FOLLOWING EXTRA WORK.

Description of Extra Work:

      Reimburse Contractor for lost labor, material and equipment for dealys resulting
      from mismarked gas mains as detailed below.

Reason for Need for Extra Work:

      A. Down-time awaiting decision on relocating 30″ sewer on Weaver Rd. at Sta. 50^+44
          resulting in addition of MH #110-A and relocation of MH #110.
      B. Revise direction of 36″ sewer out of H # 37 at Sta. 70^+55 on Muskrat Bay Rd. in
          order to avoid mismarked gas main; resulted in addition of MH #37-A.

Extra Work Order Cost: _ _ _ _ _ _ _ _ (Per Attached Itemized Breakdown) _ _ _ _ _ _ _ _ _ A. _$ 693.82
                                                                                              B. _ $2,792.91

Total Extra Work Cost: _ $3,486.73

Revised Contract Cost: _$5,603,151.51

     All work has been done to the best of our knowledge and belief in accordance with the
applicable provisions of the Contract and Specifications.

APPROVALS:

OWNER: _____ DATE: March 20, 1978.

CONTRACTOR:_____ DATE: March 20, 1978.

ENGINEER:_____ DATE: March 20, 1978.

     NOTE: Contractor would submit an itemized breakdown of costs covering labor, material,
             equipment, insurance, and overhead & profit.

FIGURE 6.27. Extra-work order.

7

7.1. Falsework

Falsework is the temporary support for holding the various units of a structure in position until they can be properly connected to each other and the structure can stand by itself. (See Figs. 7.1 and 7.2.) For multistory buildings and ones covering a large area, the falsework and forms (see Section 7.22) are designed for rapid assembly and disassembly so that they can be quickly moved from one story or section to another. The plans must be carefully studied for wedges or jacking devices in the falsework to keep the forms or units of the structure in place until the section is completed and will stand on its own. On long-span reinforced-concrete jobs, the falsework will hold the forms in a high camber before the concreting starts; then, as the concrete is poured and weight is added, the falsework will settle or "take up," and the wedges and jacks mentioned above will be used to keep the forms at the proper elevation and camber (see Section 4.3). On structural-steel jobs, the falsework will hold the units in place until the connections are all bolted or welded.

Falsework can be made of timber or metal, or a combination of the two. The timber ones will be erected by carpenters with labor helpers. The metal ones, which will be built-up sections that sit one on top of another, can be erected by labor, but some projects require that the work be done by steelworkers.

Each section must be set on a proper footing, which can be either a steel plate or timber. The timber units will have hardwood wedges under each column, while the metal ones will have screw devices similar to screw jacks.

FIGURE 7.1. Built-up falsework for a multiarched reinforced concrete bridge. The steel beams provide an opening for the passage of debris carried by the river when at flood stage, about 4 ft higher than the level shown. Note that one set of arches has been poured, and the forms are being placed for the second span. The bridge had three arch spans and beam-and girder spans at each end.

FIGURE 7.2. Sections of steel-pipe scaffolding being used as falsework to support the forms for a reinforced-concrete bridge slab. Such sections are more commonly used during building construction. Note the screw jacks at the top and bottom of the vertical pipes. These are used to adjust the forms as the weight of the concrete is placed on them.

165

7.2. Fascia Boards

Figure 24.23 shows what a fascia board is. They are attached to the ends of the rafters and are for architectural effect mostly, but sometimes the eave trough is attached to them.

7.3. Female Workers

Female workers are entering the construction work force in ever-increasing numbers, and the law says there shall be no discrimination. Whether a member of your crew is male or female should not be considered when assigning work. Over the years, with an all-male crew, some members were physically stronger than others and were assigned the harder tasks, the stronger doing the heavy work and the weaker helping them. This procedure should be followed regardless of sex.

For some years yet you may observe resentment among the males, especially the older ones, at having females enter what has been an exclusively male domain, and that resentment may take the form of verbal and/or physical harassment. This must not be tolerated. Each worker must understand that he or she is on the job for one purpose only: to carry forward the production on the project.

As has been noted before, your job is to get your assignment completed correctly as quickly and as economically as possible. To do this, you must assign each worker to the task you believe he or she can do best, regardless of sex, age, or other factors.

7.4. Fencing

Most treatment or processing plants, pumping stations, or other structures will be surrounded by fencing of some type, usually a heavy wire mesh. The wire fabric will be held up by metal posts set in concrete. Such fencing will be at or near the property line, although on large tracts it may be only around the structures. The plans will show the alignment, with details of the fence and its supports.

The alignment should be straight between corners, and the wire fabric should be taught. With a long line of fencing, there will be evenly spaced stiffener panels to keep the wire from sagging. Care must be taken to see that the posts are vertical and that they are all at the same distance above the ground. It is best to place the top of the concrete footing a few inches above the ground to keep moisture from the metal posts.

Gates through the fence will have braces or turnbuckles to keep them from sagging.

Some fence wire will be plastic covered, and if so, it must be carefully handled. Any broken covering should be repaired to prevent rusting of the wire.

7.5. Fiber Pipe (See Section 16.7)

7.6. Fill

Fill is material used to fill a void or depression. It can be almost any kind of clean material that can be compacted to prevent settlement. It can be earth, run-of-bank gravel, broken-up concrete slabs, old building blocks or bricks, cinders, crushed stone, and so on. The kind of material acceptable will be indicated in the specifications. If the fill is to change to contours for landscaping, very little compacting may be required, but if it is to support a roadway, sidewalk, or a structure of some kind, compacting in layers will be required. Many projects will require the compacting to be checked by laboratory personnel. Get instructions from your superiors on just how much compacting will be required, and see that it is attained, as later settlement can be costly.

If you are the foreman of a borrow pit from which the fill is being obtained, see that the vehicles move in, load, and move out as quickly as possible. (See Sections 4.32. and 6.6.)

7.7. Filter Beds

In areas of pervious soil, filter beds are an economical method of disposing of the effluent from sanitary facilities and processing plants. Figures 7.3 and 7.4 indicate how the filters are constructed. The linear footage of pipe is determined by the filtering capacity of the soil. There will be a sketch on the plans showing the number of lines and the length of each. Two types of pipe are used, either clay pipe, which will come in lengths of 12–18 in., or fiber pipe; in 8–10 ft lengths. The clay units will be placed with the ends about ¼ in. apart to allow the water to flow out, and there will be a piece of tar paper over the top half of the joint to keep silt from getting in. The fiber pipe will have two or three rows of ⅜-in. holes, which are placed facing downward when the pipe is laid.

As shown in Figure 7.5, the pipe is laid on a bed of stone or gravel. The grade or slope of the pipe is very important and should be exactly as called for on the plans. If the slope is too flat, most of the water will flow out in the first few feet and overload that section of the filter bed. If the slope is too steep, the water will quickly flow to the lower end and overload that section. The percent of slope is intended to allow a slow flow so that some of the water will flow out at each opening along the pipe's full length.

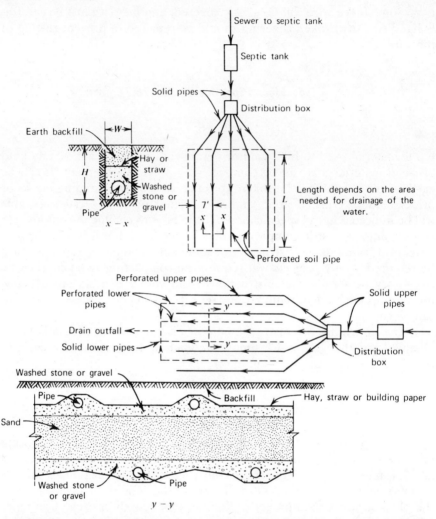

FIGURE 7.3. Filter bed. The alignment of filter-bed pipes can be done in almost any manner so long as they drain properly. However, the layout shown is the one most commonly used.

The tar paper or straw is to keep the soil from filtering down into the stone and reducing its filtering capacity.

7.8. Fire Doors

Fire doors are installed in the openings in fire walls (see Section 7.10). The doors will be made of metal or will have a fire-resistant core covered with metal. Some doors will be on a rail flush against the wall and held open

FIGURE 7.4. One length of a large sanitary drain field. The pipe is 4 in. clay with open joints set on 5 in. of crushed stone. An additional 5 in. layer of stone is being placed over the pipe. Note that the stone is being distributed by wheelbarrow from the truck in the background. The upper layer of stone will be covered with fabric or a bed of straw to keep the fine material from filtering down into the stone when the remainder of the trench is backfilled.

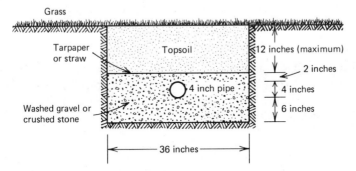

FIGURE 7.5. Filter pipe as used in a sanitary drain field. When used for draining an area, the pipe is installed near the bottom of the trench.

with a soft metal clip, which will melt quickly in case of fire and allow the door to slide closed. The other type will swing on hinges and be held open by a soft metal clip. When fire melts the clip, a weight and cable will pull the door closed. In either case, the door must fit tightly against the wall to prevent flames from getting around it.

Fire doors are usually kept closed, but since many must be open to allow movement in the facility, they will be held open by the soft metal clips noted above.

After a fire door has been installed, it should be tested several times to be sure it operates properly. You should use a match to melt the clip, but keep the flame away from the metal as the heat alone should melt it.

7.9. Fire in the Hole

The cry "fire in the hole" is the usual warning when blasting is about to occur, although a blast of a whistle or horn is sometimes used. The cry is better since there is no mistaking its meaning, while a whistle or horn could mean something else. If there is to be blasting in your work area, you should be prepared to send your workers to safety if that is necessary. Proper blasting will be confined. If it is in an open area, the blast should be covered to prevent pieces from flying through the air. However, sometimes pieces are blown out and can be carried quite a distance before falling. It is these flying pieces of which you must be aware.

7.10. Fire Walls

Fire walls are placed at intervals throughout a structure to prevent the spread of a fire throughout the building. The wall will be brick, concrete, or other fire-resistant material, and it will run from the foundations to above the roof line. All openings will have fire doors.

7.11. First Aid

On most projects of any size, the contractor will have a trained first-aid person on the site who will attend to all injured workers. However, any "up-and-coming" foreman should be interested enough in his work to take the standard Red Cross first-aid course. The coursework is not hard to master and takes only a few evenings to complete. It could save the life of one of your workers. Many injuries have been made much worse by improperly moving the victims. If you have a worker injured and you are not a qualified first-aider, get skilled help to care for the victim.

On all accidents, the foreman will be required to fill out accident forms. Most contractors will have forms supplied by their insurance company. If none is available, make out a report on a sheet of plain white paper, making a copy for yourself. Give all the facts as you know them, and do not assume conditions. Be accurate and sure of the facts, as you may be required to testify in court on what you wrote.

On minor injuries, the worker may say he is all right and want to return to work. It is best to have the worker sent to a doctor or hospital for a routine checkup. It could save costly injury claims later, if the problem proves to be worse than first thought or if the victim claims that later problems are the result of the accident.

7.12. Fish Ladders

Fish ladders provide a way for fish to get upstream to spawn. The ladder can be a series of shallow ponds, one a little higher than another, so the fish can jump from one pond to another until it gets to the top. Another type is a straight slope with several inches of water flowing, through which the fish can swim upstream.

The plans will indicate the type to be used. They will usually be made of concrete, which will be formed and poured like any other concrete.

7.13. Fixed Ends

(See Section 7.23)

All beams, girders, and trusses of any length will have a fixed end and a free end. The fixed end will be tied to the support surface, the method depending on the material of which it is made.

Steel units will have a bearing plate, with anchor bolts through the plate into the support surface. Concrete units will be poured directly on the support surface, with dowels to hold the unit in place. Wood units will usually have steel angles with anchor bolts.

The anchor bolts must be carefully set so that they will fit the holes in the units (see Section 2.12).

7.14. Flashing

Flashing is used to waterproof the joint where a chimney or vent pipe passes through the roof or where the roof connects to the wall of a higher section. Figures 7.6–7.8 show how the flashing is installed.

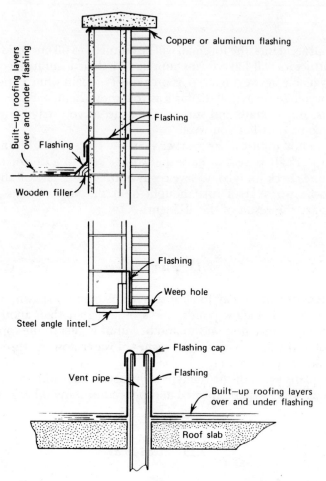

FIGURE 7.6. Various ways flashing is used to prevent leaks in buildings.

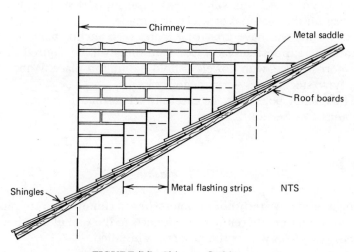

FIGURE 7.7. Chimney flashing.

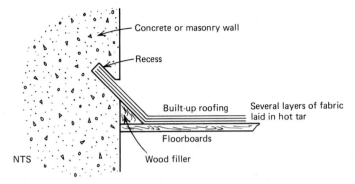

FIGURE 7.8. Wall flashing.

The metal flashing may be galvanized tin, copper, or aluminum. The strips around a chimney should fit tightly together so that snow or driving rain cannot be blown back behind the sheets. On a built-up roof, the roofing sheets should be back the full depth of the recess, and the hot tar should be wiped well back into the recess before the sheets are inserted. On vent pipes, the cap must be pinched tight so that it cannot be blown off. When flashing sheets are placed in the joint of a masonry wall, the sheet should be placed on the masonry before the mortar for the joint is laid.

7.15. Flatbeds

A flatbed is a low, heavy trailer designed to carry construction equipment such as bulldozers, backhoes, cranes, and other equipment with caterpillar treads that could damage the roadway. Some have straight beds, and others will be lower between the wheel axles. The trailer will be pulled by a tractor, the same as any tractor–trailer unit.

All equipment to be moved by such a trailer should be well secured, even for short moves. The best equipment for securing the machines is a chain and ratchet.

7.16. Floating Equipment

(See Section 5.17.)

Floating equipment is used during the construction of bridges, docks, and other structures in or over the water.

On multiple-span bridges, floating equipment is used to construct the cofferdams, piers, and falsework between the piers if there is any.

While constructing the cofferdams, the equipment is held in place by anchors at each corner of the unit. On some jobs, clusters of wood piling are driven so that the equipment can be tied to them. Once a cofferdam is completed, the equipment is tied to it.

The handling of floating equipment is affected by river currents, tides, wind storms, and ice. Large-sized used truck tires are usually hung along the side of a barge or other equipment in order to protect it when driven against an object by strong winds or river currents. There are commercial bumpers on the market, but most people use the old tires. When ice is floating in the river, it can pile up against a barge and put strong stress on the anchors. During such times, a worker should be assigned to push the ice around to the sides by use of a long pike pole.

All floating equipment will leak to some extent. The accumulation of water leaking into the hull is controlled by "bilge" pumps, either manual or power driven. It is usually the foreman's job to see that the bilge water is kept below a specified level.

Most floating equipment does not have its own power to navigate, so it must be pushed into place by a powerboat or tug. However, such equipment can be moved by taking in on one set of anchor ropes while letting out on the other. On heavy units, there will be a power windlass to assist in pulling in on the anchor ropes.

Floating equipment which has a superstructure higher above the water than the low point of a bridge downstream must have special security to prevent it from breaking its mooring and floating or being blown against the bridge. (See Figure 7.9.)

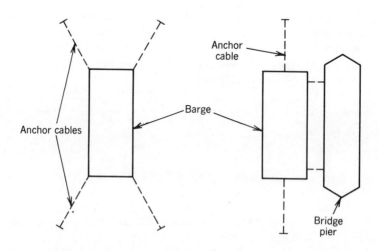

FIGURE 7.9. A method of securing floating equipment when *A* in open water and (*B*) when near a solid structure.

7.17. Floors

There is a wide range of material for floors. It can be a simple slab of reinforced or plain concrete on the ground; it may be wooden joists and floorboards; it may be reinforced-concrete beams and slabs; it can be structural-steel beams with concrete or wooden flooring; and so on. A floor system is no better than the material that supports it. For a slab on the ground, the subgrade must be in the proper condition to receive and support it (see Section 20.61.). For flooring supported above the ground, the foreman must rely on the designer to specify the correct materials. The foreman must see that all units of the floor system meet the requirements of the plans and specifications as to material, size, and quality. The various connections must be of proper size and quality. The flooring must be as specified, and special care must be taken to see that all openings are according to plans. (See Sections 3.23, 4.35, 19.52, 20.58, 21.18 and 24.13.)

Any bridging between joists or beams should not be installed until the finish flooring is in place.

Most specifications will indicate how finish flooring is to be protected until the structure is turned over to the owner. In the absence of such information, the foreman should confer with his superior, as repairs to finish flooring are a very costly operation.

In a structure with a structural-steel frame (girders, beams, and joists), the flooring may be precast units clipped to the steel, or it may be reinforced concrete poured in forms between the joists. When the floor loads are to be light, a paper fabric with steel wire mesh attached to it may be spread over the whole area and concrete will be poured on it. As the concrete is poured, the weight will cause the fabric to settle, but the wire mesh will stay in place and be embedded in the concrete. Such fabric must be well secured around the edges, or the weight of the concrete will cause it to collapse.

The finish surface may be tile set in mortar, wood nailed to strips set in the concrete, asphalt tile set in plastic, or terrazzo, or the concrete surface may be troweled smooth and painted.

7.18. Flow Diagrams

Figure 7.10 shows what a flow diagram is. The one shown is for a sewage treatment plant. It shows the relation of the various units to each other and gives the elevations of flow in each. The diagram is used during construction to see that the piping will allow flow between units as intended. After the plant is in operation, the operator will use the diagram to check flow and recycling.

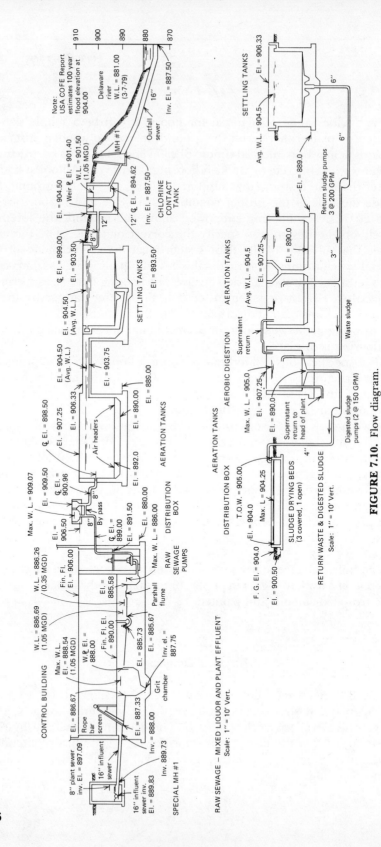

FIGURE 7.10. Flow diagram.

7.19. *Flues (See Section 4.23)*

7.20. *Footings*

The footing is the base foundation for walls, columns, piers, and abutments. No structures is better than the foundation on which it rests, but because footings are out of sight after backfilling, there is sometimes a tendency to relax on the quality of the workmanship. Such a tendency could be disastrous. Settlement due to faulty foundations can cause cracks and even failure, either of which could be costly to the contractor.

Before concrete is poured for a foundation, the subgrade should be checked to see that it conforms to the boring log and/or has been approved by the engineer. The forms, if any, should be checked for dimensions and elevations. Reinforcing steel, including the dowels, should be secured in place.

Any soft spots in the subgrade should be excavated, and the void should be filled with compacted fill or concrete.

There should be no water standing on or flowing over the subgrade.

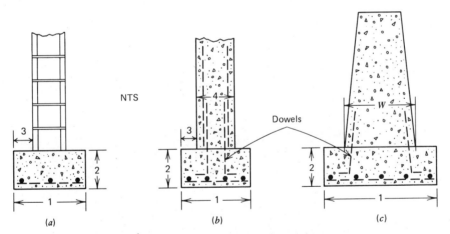

FIGURE 7.11. (*a*) Reinforced concrete footing for a concrete block wall. Dimension 1 will be from 2 to 4 ft, depending on the soil-bearing value and the height of the wall. Dimension 2 is usually 12 in., and Dimension 3 will be at least 8 in. The plans will show the dimensions required. The reinforcing bars will be ½ in. longitudinal and ¼–⅜ in. lateral to hold the long bars. The bars will be placed about 3 in. from the subgrade.

(*b*) Reinforced concrete footing and wall. Dimension 1 will be dimension 4 plus 8–12 in. Dimension 2 will be 1–4 ft, and again, the plans will give the proper figures.

A footing for a concrete pier will be similar to (*b*) except that the pier (or column) will be round or square and the footing square.

(*c*) A section of a bridge pier with tapered sides. Dimension 1 will be equal to *W* plus 2 ft or more. Dimension 2 will be 2 ft or more. Dowels are almost always used, and there may be bars in the pier as well.

If water is present, it should be directed to sump holes and pumped out. (See Sections 4.35 and 20.61.)

Figure 7.11 shows three types of footings.

7.21. Force Mains

Force mains are used to pump sewage to a higher elevation. When gravity sewers extend over long distances in flat terrain, the necessary slope would require trenching to great depths unless pumping stations and force mains are used bring the slope back up. The force main will be of a smaller pipe than the gravity-flowing main sewer. The pipe and its joints will be able to withstand the high pressures. The pipe will be laid to a constant depth below the surface, rather than to a set slope. The high points in the pipeline will have vacuum-release valves, and the low points will have pressure-release valves. The plans will show the type and location of the valves. Force mains will require thrust blocks or tie rods, the same as water mains. (See Section 3.20.)

Like any pipeline in soft ground, the force main may require a cradle. (See Figure 7.12.)

7.22. Forms (Concrete)

Forms for concrete will range from a few planks as forms for a simple footing to very elaborate units for multiarched concrete bridges. For concrete

FIGURE 7.12. A 24-in. reinforced-concrete sewer force main supported on a steel bridge over a creek. The pipe will be encased in a fiberglass insulation blanket.

units above the ground, the forms will be supported by some kind of falsework (see Section 7.1). In all cases, the forms must contain the concrete until it has hardened sufficiently to support itself. The specifications will indicate how long the forms must be left in place and which forms may be removed first.

The tightness of the forms will depend on the quality of surface finish required and the desire to eliminate honeycomb (see Section 9.9). Leaky forms will allow mortar to run out through cracks, causing honeycomb, and as the mortar sets and becomes hard, fins will form on the surface as the forms are removed. It is difficult to remove the fins without exposing the coarse aggregate.

Most forms are now held in place by patented metal devices which hold the form rigidly in place and prevent the form from moving in or out. However, wire with wood spreaders can be used. (See Figure 7.13.)

Most specifications will require the forms to be smooth on the inside and to be treated with oil, paraffin, or another substance in order to prevent the concrete from sticking.

Many contractors rent, or own, patented commercial forms. There are several kinds on the market, but all are similar. They will be secured by metal devices which hold the form rigidly in place so that it cannot move in or out. Such spreaders are designed so that they can be broken off ½– 1 in. back from the surface of the concrete. The small hole left in the concrete will be filled with mortar to prevent rust stains on the surface of the concrete.

When forms are removed, they should be cleaned immediately, oiled, and stacked in a safe place in such a manner as to prevent warping.

Study Figures 7.13 and 7.14 carefully, and will see that forms are designed so that the side forms for beams and girders, and those supporting the slabs between beams, can be removed before those supporting the beams and girders.

Most specifications will allow the side forms, and those supporting short-span slabs, to be removed in 7 days, while those supporting beams and girders will require 14 or more days.

The foreman should satisfy himself that the forms are well secured in place and are strong enough to hold the concrete in position. It is not uncommon for forms to spread when the full weight of the concrete is pressing on them. Such movement will cause misalignment of the concrete, which is difficult and expensive to correct. The forms should be cleaned of all sawdust, dirt, and other foreign matter (also snow and ice in winter) before concreting begins.

The foreman should keep a close watch on the forms as the concreting proceeds, and if any movement is detected, the pouring should stop until the forms have been realigned and reinforced.

Some plans call for mechanical or electrical units or surface covering to be attached to the surface of the concrete wall or ceiling by metal devices

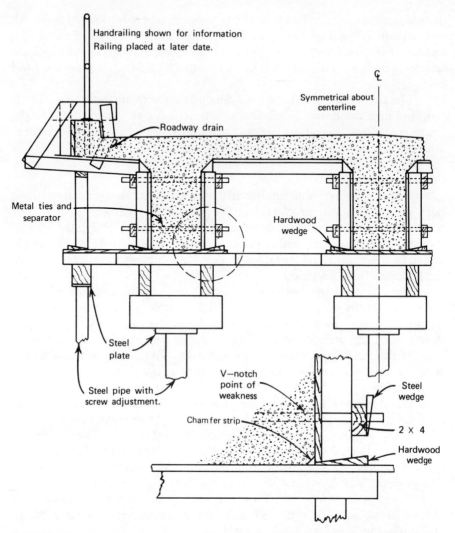

FIGURE 7.13. Concrete forms. One method of forming for the deck of a reinforced-concrete bridge. A similar design would be used for the floor of a building. (See Section 7.1.)

cast in the concrete. Such devices will be attached to the forms and remain in the concrete as the forms are removed. (See Figure 7.13.)

The plans may call for ridges, grooves, or other architectural features in the surface of a concrete wall. These are obtained by devices attached to the forms, the same as noted above. When the forms are removed, the devices come off, with the forms leaving the desired design in the surface of the wall.

All sharp edges and corners of concrete should be blunted by using chamfer strips (see Figure 7.13). Such sharp edges and corners are easily

FIGURE 7.14. Crew erecting a section of exterior forms for a circular tank. This wall is for one of the base slabs shown in Figs. 4.21 and 4.23. Note that reinforcing bars were placed after the interior forms were erected. The wall will be poured in sections with waterproof construction joints between. Note the steel pipe bracing for the interior forms.

damaged, and a break in the surface will expose the large aggregate and be difficult to repair.

The forms will be constructed and erected by the carpenters with help from labor crews.

7.23. Free (Expansion) Ends

(See Section 7.13).

The free end of a beam or girder allows for expansion and contraction. There will be some kind of device to allow the beam to slide as it expands or contracts. The plans will show the type of device, usually a heavy steel or bronze plate with the top plate secured to the beam and the bottom one secured to the support surface. Sometimes it is required that graphite be placed between the plates.

The centerline of the plates must be parallel with the centerline of the beam or girder so there will be no binding as the beam moves.

7.24. Freezing

Frost and freezing temperatures can cause problems on civil engineering projects. concrete, masonry, subgrade work, painting, and even handling structural steel can be affected by frost and freezing weather.

For concrete, there must be no frost on the aggregate, and the water should be at least 75 deg F before it is put into the mixer. There must be no frost on the forms or the reinforcing steel as the concrete is poured. After the concrete is poured, it must be kept above freezing for 14–28 days. The specifications will indicate the time required.

Most projects today will receive the concrete from a central Redi-mix plant, and it will be heated to a required temperature. The area of the forms will usually be enclosed by tarpaulins or plastic sheeting with heating underneath to remove the frost before pouring and to protect the concrete after pouring.

Masonry is protected from freezing in much the same way as concrete. The masonry units and the mortar will be heated, and the area will be enclosed and heated. (See Figures 4.25 and 4.26.)

Whenever possible, fine grading a subgrade should not be attempted if there is danger of freezing weather. However, the progress of the job may require it. If the frost is only on the surface, it can be eliminated by spraying gasoline lightly over the surface and setting it on fire. If the weather will not be too cold, the surface can be protected by covering it with hay or straw about 12 in. thick and covering that with tarpaulins or plastic sheets. Tarpaulins or plastic sheets can be placed on a wooden frame 12 in. or so above the surface, and heat can be forced into the enclosure.

Painting should not be attempted if there is frost on the surface. In fact, painting should not be undertaken until the air temperature is at least 40 deg F and rising.

When structural steel is covered with frost, it becomes very slippery and is dangerous to handle or walk on. If the sun is shining on the steel, it will quickly remove the frost. If not, it can be removed with a kerosene torch.

7.25. French Drains

French drains (farm drains)) are used to dry up swampy areas or areas with high groundwater (see Figure 7.15). For many years, farmers have used these drains to dry up soggy fields so that they can be cultivated.

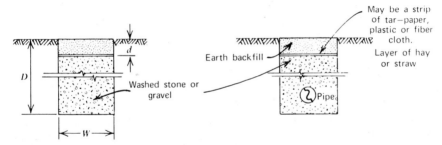

FIGURE 7.15. French drain. The dimensions of a French drain can be whatever the engineer specifies. The depth of the cover must be specified by the health department if contaminated water is present. Otherwise, it should be at least 12 in. A French drain may or may not have a small-diameter pipe to speed up the flow away from the area. The drain must slope to a lower area to carry off the water.

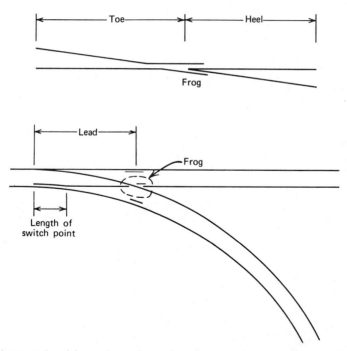

FIGURE 7.16. Railroad frog. The position of the frog at the junction of two tracks and the position of the moveable switch points are shown.

Such drains are used along the edges of highway pavements to drain water from the subgrade and prevent frost heave. They are used around tennis courts and athletic fields for the same purpose.

A system of French drains must slope to drain the water to a point of disposal. A point of disposal could be a creek or a low area where ponding is permitted, or an outfall pipe could carry the water to a point of disposal.

Care must be taken to see that the barrier on top of the stone will prevent silt from being washed down into the voids between the stones and thus reducing the effectiveness of the drain. In silty soils, the trench for the drain is sometimes lined with fabric to prevent the silt from entering.

7.26. Frogs (Railroad)

Figure 7.16 shows how a frog keeps the wheels of a railroad car on the right rails when the car is switched to a sidetrack. The way the switch points

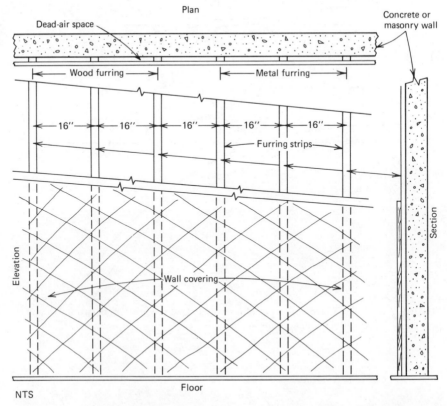

FIGURE 7.17. Furring strips for installing wall covering over concrete or masonry walls.

are set in the figure, the wheels will be guided onto the sidetrack. If switch points were moved down (in the figure), the wheels would stay on the main rails.

7.27. Frost Heave

Frost heave during construction can be of concern for several reasons. Moisture in the ground can freeze and expand and move control stakes off line and grade. Weep holes in brick walls can become plugged, allowing moisture to collect and freeze and push the bricks out of line. Frost in a subgrade can raise the surface an inch or more.

7.28. Furring

Furring is wood or metal strips attached to a wall or ceiling to which sheetrock or other covering is nailed or clipped. (See Figure 7.17.) The strips are attached by masonry nails. The strips must be attached vertically and at the proper centers to fit the sheetrock or other covering. After the wall covering is installed, the strips provide a dead-air space, which provides insulation.

8

8.1. Gabions

Gabions are used to prevent erosion and to contain an embankment or fill. Galvanized wire baskets filled with stone are placed like masonry blocks. The wire baskets are about 2×2 and 4 ft long. When starting the installation, the first row of baskets is set in place with the tops open, the baskets are filled with stone from 2 to 8 in. in size, and the tops are then closed and wired shut. The next row is placed on top of the first, filled with stone, and wired as before. This operation continues until the desired height and length are obtained. On extreme heights, two rows will be used. The wall of baskets may be vertical or slant back toward the fill. (See Figure 8.1.)

8.2. Gain Water

When concrete is deposited in depths of several feet, as in a wall or large footing, the weight of the aggregate as it settles forces the free water up to the surface. If this water is allowed to accumulate on the surface of the fresh concrete, any new concrete added will have its strength reduced by the extra water. The water accumulation can be removed by drilling a few holes in the forms and letting the water drain out. When the concrete pour is finished, any gain water on the surface should be removed, as it will form laitance when dried up. (See Section 13.4.)

FIGURE 8.1. A gabion installation to prevent erosion of a highway embankment. The galvanized wire baskets are still in good shape, although they have been in place for many years.

8.3. Gallons Per Minute

Gallons per minute (gpm) is a measure of the flow of liquid. The capacity of pumps is rated by the gpm produced at certain rpm. (See Section 18.21.) Usually, there will be a chart with the pump listing the gpm flow at various "heads" or heights of the liquid above the pump.

8.4. Gaskets

Many of the leaks in a newly installed pipeline are due to careless installation of the gaskets and can mean a costly repair job for the contractor. As pipe sections are being connected together, the ends and the gaskets must be clean and lubricated, and the connection must be made carefully so as not to displace the gasket.

The pipelayer should have a metal gauge with which to check the gasket after the joint has been closed. The gauge will tell if the gasket is at the proper distance from the end of the pipe. If the gauge loses contact with the gasket as it is run around the pipe, the gasket has been displaced and the joint will probably leak. The pipe should be disconnected, and the gasket should be reset, as it is much less costly to do that than to dig it up later if a leak develops.

Undetected displaced gaskets can allow groundwater to carry silt into a sewer line, causing a void around the pipe. Such a void can allow the pipe to settle, which can break the pipe or open a joint, either of which can allow more silt to enter. The void can develop to a size that will cause the ground above to settle under a roadway or structure. Too much silt can also damage the sewage pumps. Gasket leaks in water mains can flood nearby structures or undermine a road.

8.5. Gate Valves

Gate valves are used in pipelines when it is necessary to shut off completely the flow of water or other liquid. The gate within the valve housing slides up and down to open and close. The gate will fit into grooves with a tight fit for a complete seal. Because of these grooves, each valve should be checked before installation to see that no grit or other foreign matter has lodged in the groove, as it will prevent the gate from closing fully.

8.6. Gels (Jelly)

Chemicals that form a semisolid mass when injected into the soil are used for excluding water or moisture by two methods.

One method is to inject the chemical into the soil along the line of proposed trenching. The chemical forms a gel that stabilizes the soil so that the trench banks will not collapse. It is also injected into the soil along the foundation of a structure to keep groundwater from penetrating the wall.

The other method is used in sewers to seal leaks through cracks in the pipe or through leaky joints. (See Figure 21.22.) The method is used in conjunction with TV inspection (see Section 21.36).

8.7. General Foremen

Large projects employing many workers may have a general foreman to assist the superintendent, who is often tied up with administrative matters. The regular foremen will report to the general foreman.

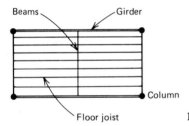

Beams
Girder
Column
Floor joist

FIGURE 8.2. Structural units in a typical floor system.

8.8. Girders

A girder is the structural member that supports the smaller beams, which in turn support the still smaller joists. (See Figure 8.2.) It may be of timber, structural steel, or reinforced concrete.

8.9. Glazing

Glazing is the installation of glass in the frames of windows, doors, and other wall openings. The glass comes in various strengths, depending on where it is to be used. In wooden frames, the sheet of glass is secured by small metal wedges and putty applied around the edge to seal the joint. In metal frames, the glass sheet is held in place by metal strips with rubber or vinyl gaskets. This work is usually done by glazers. If you are involved in glass installation, be sure that the glass sheet fits closely in the frame, that it bears tightly against the frame, and that the metal wedges are installed closely enough to hold the glass in place. The putty should be applied by a putty gun, which will make a smooth run of constant width.

8.10. Glide Angle

The glide angle is the minimum angle at which an airplane can glide from a given elevation and reach the runway. The glide angle also shows the minimum angle at which an airplane can fly and clear trees, building, and other objects beyond the runway when taking off. Most plans for airfields show the glide angle; all obstacles must be kept below that elevation.

8.11. Grade Beams

Referring to Figure 8.3, a grade beam will be at or near ground level. It can be a perimeter beam under only the exterior walls, or it can be a system of beams and girders supporting the ground-level floor. If the beam is

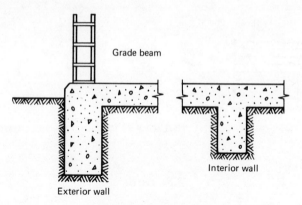

Grade beam

Interior wall

Exterior wall

FIGURE 8.3. Grade beam.

underground, it will usually be made of concrete. If it is at or above ground level, it may be made of structural steel or timber.

8.12. Grade Line and Pole

Refer to Figure 2.4. A mason's line or wire stretched taut between two batter boards will maintain the grade between them. So a grade pole marked with the distance from the batter board to the invert of the pipe will give the proper grade for the pipe at any point between the batter boards.

8.13. Grades

Grade can mean either the degree of slope from horizontal for roadways, railroads, sewers, or landscaping or the quality of various materials such as lumber, aggregate, and other materials.

The grade of a slope will be given by a percentage. A 3% (.03) grade for a roadway means the surface will rise (a plus grade) or fall (a minus grade) 3 ft in 100 ft.

The grade of material can be A, B, and so on, or #1, #2, and so on. Usually, the "A" is the highest quality, such as with grade-A lumber. In aggregate, the #1A is the smallest.

8.14. Grade Stakes

The surveyors will set grade stakes indicating the line and grade of the work. They will establish the limits both horizontal and vertical, for the main parts of the structure. There will be a tack in the top of the stake for

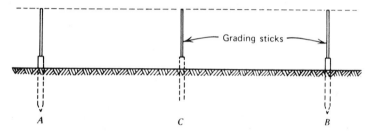

FIGURE 8.4. Replacing grade stakes using grading sticks.

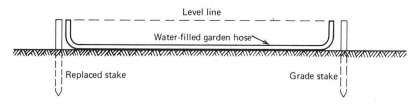

FIGURE 8.5. Replacing grade stakes using a garden hose.

reference to horizontal lines, and the top of the stake will have an elevation established on it for reference to vertical lines or elevations.

If you are to be involved in the use of grade stakes, you must know what each stake represents and be assured that the stakes have not mixed since they were placed. As work proceeds, be sure none of your crew or equipment disturbs the stakes. (See Sections 2.10, 8.15, and 13.12.) (See Figures 8.4–8.6.)

8.15. Grading (Earth)

Grading is the rearranging of the earth's surface by excavating the soil at one place and placing it at another. The excavating is called "cut," and the placing is called "fill."

During the layout of a subdivision, the tops of the higher hills are cut off and placed in the lower valleys to provide gentler slopes. The grading will be controlled by grade stakes placed about 50 ft apart in a grid pattern. The stakes will be marked for "cut" or "fill," as the case may be. (See Section 20.45.) The amount of cut or fill marked on the stakes will usually refer to the finished grade, including the topsoil, if any, so the depth of the topsoil must be taken into consideration when placing the earth. (See Section 21.21.)

Grading for a highway will involve cuts through the hills and filling in the valleys. When the grade of the road is planned, engineers try to set the grade so that the cut and fill will balance as nearly as possible.

Date _5/31/77_
Contract _Minetto S.D._
Job No. _138.25 Contr 1-B_

Grade Letter

IN/S N.Y.S. Route 48 From _3+90 (MH 57_ To _8+01 / MH 58) P_
Street   Station   Station

_8_ In _____ Pipe
Size   Type

Manhole No.	Station	Percent Grade	Inert Elev.	Offset Stake			Center Line	
				Offset	Elev.	Cut	Elev.	Cut
M.H 57	3+90'		321.39	15'L	336.53	15.14	334.5	13.1
	4+00		321.43				335.2	13.8
	4+40		321.59				336.1	14.5
	4+80		321.75				337.1	15.3
	5+20		321.91				336.4	14.5
	5+60		322.07				338.1	16.0
	6+00	.002	322.23	20'L	336.87	14.64	336.6	14.4
	6+40		322.39				335.7	13.3
	6+80		322.55				335.4	12.8
	7+20		322.71				335.2	12.5
	7+60		322.87				334.5	11.6
	8+00		323.03				333.9	10.9
M.H 58	8+01		323.03	20'L	333.52	10.49	333.9	10.9

Instructions.

     Field notes 5/27/77

     Plan profile sheet 11 rev. #5

     MH 57 8" sewer @ invert 321.08

Remarks: Contractor must have complied with construction procedures before additional grade letters will be issued.

To: _____ _____
     Contractor's Agent   Date   By: _GAS_   _FK_

FIGURE 8.6. Grade letter. Illustrated is a grade letter from an actual project. The centerline elevations are taken from the top of the offset grade stakes.

When the areas of cut and fill are reasonably close together (1000–1500 ft), the large scrapers will be used. For longer distances, bottom- or rear-dumping trucks are usually more economical. In either case, bulldozers will be used to level the soil, and the various types of rollers will compact it. (See Section 19.17.) The trucks will be loaded by power shovel, front-end loader, or conveyor-belt loader. (See Section 6.9.)

8.16. Gratings

Grating is an open-faced surface used for wallways, catch basins, and openings in structures. It may be made of steel bars or rods, or for catch basins it may be of cast iron. The weight of the metal will depend on the strength required.

8.17. Grievances

Grievances among the workers can be a problem and must be handled diplomatically. When a worker complains to you, listen carefully and be sure you fully understand the problem. If it is a situation you can correct, do so, as harmony among the crew is necessary for progress in the job. If the complaint is justified but beyond your ability to correct, confer with your superior. Your willingness to try to correct problems will enhance your standing with the crew. Sometimes you will have a chronic complainer who will grumble about almost anything. Such a worker can disrupt the whole crew and should be dealt with promptly. If you cannot talk him into changing his ways, recommend to your superior that he be transferred or discharged, and explain why. The nature of construction is such that it is often uncomfortable, either too hot, too cold, or too wet. You should do what you can to make the work as pleasant as possible, but you cannot control the weather. Experienced workers understand this and accept the conditions.

8.18. Groundwater

Groundwater is the natural water in the soil and is the source of water for wells, springs, and artesian wells. In areas of high groundwater, there will be problems for excavations that extend below the water elevation. There are several methods of dealing with the problem, such as by well points, by using open caissons around the area and where the ground slopes, or by trenching with French drains to carry the water to a lower elevation. Sometimes the specifications will indicate the method to be used. (See Sections 3.22, 4.1, and 24.14.)

If you are involved with excavating in an area of high groundwater, be alert for soft spots or small boils, and if any appear, confer with your superior.

Sometimes small pockets of groundwater will be encountered which will cause soft spots. These can be handled by excavating the area and refilling it with dry, stable material well compacted in place.

You should be aware that lowering the groundwater can cause settlement in nearby structures.

If you are pumping out large quantities of groundwater, you must be concerned about where it is being discharged and avoid flooding a street or some other property. If the water bears silt, this too can cause problems.

8.19. Grouting

Grouting is usually done with a mixture of Portland cement and sand, with enough water to make the type of mix desired; however, chemicals are also used for certain kinds of grouting.

Cement and sand grout is used to fill the voids between stones to make a solid mass and to fill voids under structures, under column bearing plates, behind tunnel liner plates, and under the bases of machinery. It is also used to repair the surface of spalled concrete. When used between stones, it will be thin and runny so that it can flow in between the stones and fill all the voids. It will also be thin and runny when used behind linear plates. When used under structures or bearing plates, it will be quite stiff and may have an additive to keep it from shrinking. Grout is used under pressure to raise sunken concrete slabs where a special machine mixes the grout and pumps it through a hose to holes drilled in the surface of the slab. The grouting continues until the pressure raises the slab to the required position.

Repairing spalled concrete surfaces is similar to applying stucco. The surface is roughened, and if the grout will be over ½ in. in thickness, a layer of wire or expanded metal will be attached. The grout will be applied in about ¼ in. layers, with each layer being allowed to dry before the next layer is applied. Each layer will be roughened to improve the bonding.

Chemical grouting is used to seal leaks from cracks in sewer pipes and to waterproof the outside of basement walls. The chemical is injected into the ground where the moisture turns it into a gel. In sewers, the chemical is forced out through the crack into the soil.

Contractors with special equipment and training can seal very narrow cracks in concrete slabs by injecting special chemicals which fill the crack and exclude moisture.

8.20. Grubbing

Most contracts for civil engineering work will have a pay item for "clearing and grubbing." (See Section 4.28.) The grubbing item includes the removal of trees, bushes, and other landscape items. It may include the removal of topsoil, or that may be a separate pay item. Grubbing includes the removal of stumps and roots and will include its removal off site or to a designated spot on site. Burning or burial may be specified. If you are involved with grubbing, you should inquire if you must remove all roots or only the tap root. If you are burning the debris, be sure no sparks are carried to where they could cause fires. It is always good to have water or a fire extinguisher available.

8.21. Gunite

Gunite is a grout mixture of Portland cement, sand, and water that is applied under pressure to a concrete surface or to wire mesh around structural steel for fire protection. The work is done with special equipment and work crews. The equipment mixes the grout and forces it through a heavy hose to a special nozzle where it is mixed with high-pressure air that sprays the mix against the concrete surface or onto the wire mesh.

For fire protection of structural steel, a wire mesh is attached to the steel by wires, clips, or welding. The Gunite is then sprayed on, forming a coating of ¼ in. or more in thickness, depending on the requirements.

To repair spalled or otherwise damaged concrete surfaces, the area is first cleaned by sandblasting, and if a heavy coat is to be applied, a wire

FIGURE 8.7. The Gunite repairing of an old concrete reservoir. The top half of the slope has been completed, and work is progressing on the lower half. Note the reinforcing mesh attached to the surface before the Gunite is applied. The timbers provide a platform for the workers on the upper half of the run. Note the overflow chamber at the extreme left-hand side. After the slope has been completed, the floor will be treated the same way.

FIGURE 8.8. The Gunite repair of a spalled surface of a bridge. The dark area shows where Gunite was applied after the spalled surface had been cleaned of all loose material and a wire mesh was attached. The area where the Gunite was applied will bleach out in a few weeks to nearly match the original surface. (See Section 20.40.)

FIGURE 8.9. Closeup of a worker applying Gunite. This worker should be wearing a protective face mask or safety goggles.

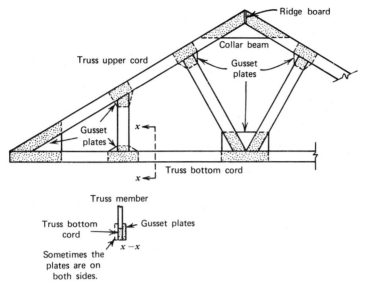

FIGURE 8.10. Gusset plates.

mesh will be attached. The Gunite is blasted onto the surface to the required thickness and allowed to set or harden. The surface must be protected from heat or cold the same as any concrete or mortar. In hot weather, the surface is usually sprayed with water or covered with damp fabric to slow the drying.

There is considerable rebound of the sand particles, so protective clothing and face shields should be used. (See Figures 8.7–8.9.)

8.22. Gusset Plates

Figure 8.10 indicates how gusset plates are used. On a timber truss, the plates can be of plywood or metal, held in place by nails or bolts. For structural-steel trusses, the plates will be steel bolted in place. The plates may be on one or both sides, depending on the type of truss and the loads it must sustain.

8.23. Gutters

Gutters receive the rainwater flowing off the surface of a roadway and convey it to a point of disposal, such as a catch basin or a ditch. (See Figures 4.42 and 4.43.)

9

9.1. Hammers

Hammers run from the light carpenter's hammer to the pile driver that weighs several tons. There are mechanical hammers that are electric or air driven, are held by hand, and use a chisel or star drill for chipping or drilling small holes. Jackhammers are used to break up old paving, rock, or other hard material and are usually powered by compressed air, although some are electric. There are several types of pile drivers. Most use compressed air, some are electric, and some use vibration to drive the piles. (See Section 17.15.)

The vibrating hammer grips the piling and vibrates it as the weight of the hammer forces the piling down. The other hammers are single acting or double acting. The single-acting hammer has a weight of 5000 lb or more, and is raised by the air pressure and allowed to drop free from about 3 ft. The double-acting hammer has a piston which is forced up and down by the air pressure. One of the most popular hammers is the Vulcan No. 1, which is single acting, with a weight of 5000 lb and falling 3 ft.

These hammers, like all tools, must be kept clean and properly lubricated. Electric tools should have a ground wire. The compressed air units should have a lubricating device in the air-supply hose. The exhaust ports in the air units must be kept clean, and the weight shaft must be lubricated.

9.2. Hand Levels

The hand level is a very useful tool for checking elevations, particularly on earthwork projects. They are a little difficult to get used to, but with some practice you will find them handy. When you sight with them, grip the level firmly but not too tightly. Keep your upper arm against your body to steady the instrument. The hand level will not give elevations accurately enough to set forms, structural shapes, or pipes, but will give close enough checks on earth excavations and embankments.

9.3. Hard Hats

Most civil engineering projects will require the use of hard hats and will have signs around the site warning that they must be used. The hats are made of metal or plastic, and while they will not protect against heavy blows, they lessen the impact and guard against severe injury from falling objects, and they protect you from bumping your head on low structures. The hats may be uncomfortable on very hot days, but you and your crew should make a practice of wearing them at all times when on the job site.

9.4. Hardpan

When excavating in soil with high sand and gravel content, you may encounter an area where the particles are cemented together. This condition is called hardpan. The degree of hardness varies from mild to as hard as concrete. If lightly cemented, it can be excavated by backhoe or dozer; however, some will require a jackhammer or blasting. When encountered in gravel pits, it is usually left by working around it. Hardpan can make a good foundation if there is a large quantity of it.

9.5. Hardware

Hardware includes such things as doorknobs, locks, hinges (butts), door closers, panic bars, kick plates, mail slots, and nails and screws of all sizes, all of which will be installed by the carpenters or will come installed in package units. If your crew is working near units with the hardware installed, be careful to see that no damage occurs, as any replacement can be expensive.

9.6. Heading

During tunnel work you will hear the word "heading" quite often. It refers to the work area where the excavating is being done. This is the most dangerous area in tunnel work. There is always the possibility of uncovering a vein of water under pressure, a gas pocket, a vein of soft mud, and, in rock areas, slabs falling from the ceiling. Sometimes the workers get careless and excavate too far ahead of the safety installations such as liner plates, cribbing, or, in rock, rock bolts. Sometimes they are slow in getting the ventilating pipe installed close to the heading, and the air gets foul.

9.7. Heating (Temporary)

Temporary heating during cold-weather construction is an overhead cost to the contractor, so some skimp on it. This is false economy. There are several methods of providing low-cost heating on construction projects. In open areas, the old 55 gal steel drum standing on end, open at the top, with a few holes punched through it near the bottom, and fueled with lumber scraps or cut-up trees which were removed during clearing will radiate a great deal of heat. The 55 gal steel drum with both ends cut out and laid end to end with a kerosene burner at one end will provide plenty of heat. (See Section 4.43.) Standard kerosene heaters can be used in enclosed areas, but you must be careful to ventilate the gases.

You must be careful at all open fires to keep sparks from falling on flammable material. It is always good to have a fire extinguisher handy. Do not throw sticks into an open fire; lay them in gently to prevent sparks from flying. (See Figure 4.26.)

Large space heaters fired by oil, gas, or gasoline may be used for temporary heat in buildings.

9.8. Heating (Equipment)

The permanent heating equipment will be installed by the plumber or the "H–V–A/C" people; (heating, ventilating, air conditioning). Labor crews are sometimes assigned to assist in this work. Uncrating the equipment and moving it to the point of installation requires care, so as not to damage any exterior controls or piping. Any instructions, tools, lubricants, gaskets, and other nonattached items must be collected and delivered to the installers.

9.9. Honeycomb

Honeycomb refers to a condition in concrete where the mortar (sand and cement) has drained away from the coarse aggregate, leaving a void. The honeycomb can be caused by the mortar leaking out through cracks in the forms, by overvibrating so that the mortar is drawn to the vibrator and away from the large aggregate, or by improper mixing. Most specifications will require that all honeycomb be cut back to sound concrete and that the void be filled with proper concrete of the same mix. This is an expense to the contractor and should be avoided as much as possible. On large projects,

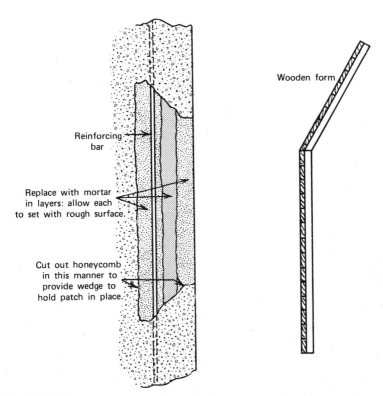

Reinforcing bar

Wooden form

Replace with mortar
in layers: allow each
to set with rough surface.

Cut out honeycomb
in this manner to
provide wedge to
hold patch in place.

FIGURE 9.1. Refilling void left when honeycomb was cut out. If there is a small void, fill it with mortar in layers. If the honeycomb is only on the surface, it is not necessary to cut down to the reinforcing. If there are large voids, install a form and pour concrete of the same mix as the original. Trim excess from the face of the structure.

FIGURE 9.2. Honeycomb on the side of the concrete bridge pier. The condition was caused by a lack of tight caulking of the joints between the form planks. The mortar leaked out, leaving the coarse aggregate exposed. The affected areas were cut back to sound concrete, and the void was filled with concrete.

some honeycomb is bound to happen, and it should be cut out as soon as possible after the forms are removed and before the concrete gets its full hardness. If the honeycomb is only near the surface, the area can be repaired by chipping off the material and filling the void with layers of stiff mortar. No layer should be over ½ in. thick, and each layer should be allowed to set before the next is applied. When complete, the area should be kept damp for several days. Deeper areas should be treated as shown in Figure 9.1. (See also Figure 9.2.)

9.10. House Connections

When sewer or water lines are installed in the street, the contractor is usually required to install the laterals up to the property line. The plumber will install the lines from the property line into the house or other structure. (See Figure 13.3)

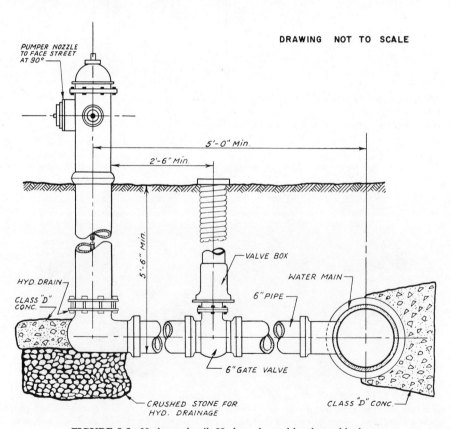

FIGURE 9.3. Hydrant detail. Hydrant braced by thrust blocks.

DRAWING NOT TO SCALE

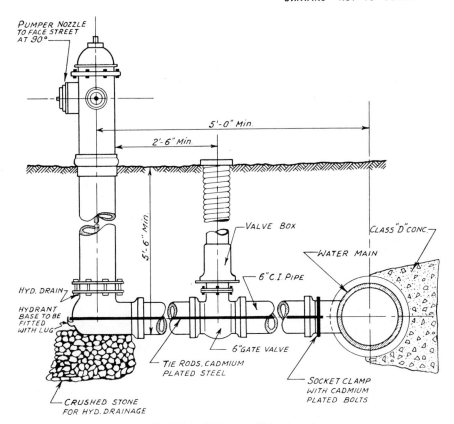

FIGURE 9.4. Hydrant held by tie rods.

9.11. Hydrants

The plans will usually show where the hydrants are to be installed and how they are to be connected to the water main. (See Figures 9.3 and 9.4.) There are two basic ways to join the hydrant to the main. The figures will show you how. If the concrete thrust blocks are used, the concrete must be set (at least seven days) before pressure is put on the line. With tie rods, the pipe sections must be up tight at all joints, and nuts on the rods must be fully tightened. If the hydrants are stored on the ground for a time before installation, you must be sure there are no small stones, sticks, or other debris inside which could get under the valve and damage it when you try to close it. The concrete for thrust blocks must be poured against solid earth so as not to give when pressure is applied to the line.

9.12. Hydration

When water is added to dry cement, a chemical reaction takes place that is called hydration. The cement and water form a paste, which when mixed with aggregate (sand and stone), becomes concrete. The strength of the concrete depends on the strength of the paste and the durability of the aggregate. Like any paste, the more liquid you add, the weaker the paste becomes. (See Section 4.35.)

10

10.1. Ice

In winter weather, ice can build up on the ground or exposed floor surfaces, making it difficult to move workers or equipment. Ice or frost must be removed from concrete before new concrete is poured upon it and must also be removed from a subgrade before work proceeds. Before concrete can be mixed on the site, all ice and frost must be removed from the aggregate. Frost on structural steel can make it difficult to handle.

Blocks of ice floating down a river can damage or destroy cofferdams, falsework, piling, and floating equipment moored in the river. Ice floes can jam in narrow points in the river or against islands and dam up the water, causing flooding over low areas along the river.

Falsework for a bridge over a river must have open bents through which the ice can be directed. (See Figure 7.1.) When ice floes start to move, workers with long pike poles should be placed along the upstream side the guide the ice blocks through the openings. In some areas, compressed air is used to prevent ice from forming in or around cofferdams or around piers. Pipe or heavy hose with small openings along its length is placed on the river bottom, and as air is applied, bubbles are formed that rise to the surface and prevent ice from forming.

Ice or frost around the job site can be removed by kerosene torches or steam. The steam is best, as it is not a fire hazard. Steam from a jenny can be directed through a small pipe to melt the ice or frost quickly (see Section 20.50).

When the temperature hovers just above freezing during the daytime and drops just below after the sun goes down, it can produce a wet surface in the day and thin ice at night. Coarse, sharp sand spread over the area can provide a nonslip footing if it does not get too cold.

10.2. Industrial Buildings

There is a number of lightweight structural-steel buildings of various designs on the market. Each patented under its own name. These structures are classified as "industrial buildings" and are usually purchased from a contractor, who furnishes and erects the building.

10.3. Infiltration

When underground piping is installed below the groundwater level, infiltration can result from leaky joints, cracks, or imperfections in the wall of the pipe. When part of a structure is below the groundwater level, infiltration can occur through poor joints in masonry, through porous masonry units, from voids around pipes passing through the wall, and from honeycomb in concrete.

Infiltration in sewer pipes can allow soil particles to pass through openings and be carried to the treatment plant where it can damage the pumps. Enough silt can plug the pipe and an excess of water can exceed the pump capacity. Plugged sewers can cause the sewage to back up into basements. A large quantity of soil passing into the sewer will cause a void around the outside of the pipe and cause it to settle, doing further damage, even collapsing the sewer.

Infiltration can be stopped in sewers with small cracks or slight leaks at the joints by injecting a chemical through the crack out into the soil. The moisture in the soil will cause the chemical to become a stiff gel capable of stopping the flow. There is special equipment designed to inject the chemicals. (See Section 21.36.)

Large breaks in sewer pipe will have to be repaired by uncovering and replacing the defective pipe. Leaks in the walls of buildings or in manholes can be prevented by applying a seal coat of hot asphalt before the area is backfilled. Sometimes the specifications will call for a waterproofing of a coat of hot asphalt, a layer of fabric, and another coat of asphalt. Once leaks occur in a wall, they can be sealed by various methods. One effective method is the following way:

1. Clean the surface of the wall in the area of the leak.
2. Roughen the surface if it is not already rough.
3. Apply a thin coat of rather dry mortar, forcing the material well into the surface to help bonding to the old concrete.
4. Leave uncovered the area where the most infiltration occurs.

5. Allow the thin layer of mortar to set after its surface has been scratched.
6. After the mortar sets, apply additional layers until the surface is again flush with the rest of the walls.
7. In the area of maximum infiltration that was left uncovered, drill a hole at the point of maximum flow and insert a pipe of the smallest diameter that will carry the flow.
8. Complete the plastering of the area, leaving only the pipe to carry the flow.
9. Keep the area damp for several days until the mortar has completely hardened.
10. After the mortar has completely set, rotate the pipe to loosen it and pull it out.
11. The hole that is left can be plugged by use of a product named Sika, which is mixed into a ball, pressed into the hole, and held there for the few minutes necessary for it to set and seal the hole. If the water is flowing with some pressure, a dry, soft wooden plug should be driven into the hole to at least 2 in. from the surface. The plug will stop or reduce the flow so that the Sika can be applied. (See Section 20.25.)

10.4. Injuries

Injuries will happen on even the best regulated jobs. They are hard on the workers and expensive to the contractor. (See Section 2.3.) If one of your workers receives an injury, you should report it to your superior at once and to the first-aid person if there is one. A seriously injured person should not be moved until trained personnel arrive. He should be made as comfortable as possible and covered if a blanket is available. Most contractors will have an accident report form which may be given to you, but you should not attempt to describe an injury unless you are trained to do so. The injury may result in court action, and the outcome may rest on the description of the injury.

10.5. Inserts (Concrete)

Inserts, usually of cast iron, are sometimes cast in concrete walls, floors, or ceilings for attaching wall coverings, rods for drop ceilings, or permanent equipment. The plans will show the type and location of the inserts, and care must be taken to see that each item is placed exactly as required, or else the attachments might not fit. The units must be secured in the forms so that the pouring of the concrete or the vibrating will not dislodge them. Correcting inserts after the concrete has set is a difficult and costly operation. The concrete must be well vibrated around each unit to hold it firmly in place when the weight is placed on it.

10.6. Inspectors (Resident Engineers)

The construction inspector is the owner's representative closest to the work. He will usually work under and report to a resident engineer, although he may report directly to the owner. His responsibility is to see that all work is done in compliance with the plans and specifications. He does not have to know how to carry on the operations to complete the work, but must know if it will comply with the plans and specifications when done. He has no right to tell you how to do your job, but may tell you that your methods may produce a result that will not be acceptable. In such a case, confer with your superior at once. You should try to be friendly, but not a buddy, with the inspector. Any instructions or criticisms of the work must be given to you, not to members of your crew. If you are away from the site temporarily and he believes a worker is doing something that will not be acceptable, he may stop him but he must inform you as soon as possible. If the inspector believes your crew is working in a manner that will result in unacceptable work, he may order you to stop. In such a case, talk the matter over with him and try to reach an agreement on how to proceed. If you cannot agree and he insists on stopping, the order should be in writing, detailing why it is necessary. (See Section 20.55.)

10.7. Instructions

You must be absolutely sure you understand all instructions received from your superiors. Do not hesitate to ask if you do not fully understand. Improper work will be an embarrassment to you and may be costly to your employer. Your request for clarification will indicate that you are truly interested in your work and want to do it right the first time. Likewise, your instructions to your crew must be fully understood by each member. Encourage your workers to ask questions if they do not fully understand.

Written instructions will come with each item of permanent equipment to be installed on the project. These must be preserved and turned over to your superiors. (See Section 6.10.)

10.8. Insulation

There are several types of insulation, such as rigid planks, batts, and loose and foam insulation. A dead-air space between masonry blocks and the brick facing of walls provides insulation. Rigid planks, 2 in. thick or more, are used on roofs and inside masonry or concrete foundation walls. Batts are used between the studs of frame buildings and between ceiling joists. Loose insulation is used between ceiling joists and sometimes between the studs. Foam insulation is usually pumped into existing frame walls between the studs.

Rigid planks are nailed or clipped to the roof rafters and placed against the inside of foundation walls before the backfilling is done. The batts will be nailed to the studs with a special nailing tool. When used on the ceiling, they are just placed between the joists. The loose insulation is blown into the ceiling between the joists and between the wall studs of a completed wall by cutting openings through the exterior covering. The opening is closed after the space is full. Foam is pumped into a finished frame wall by cutting openings, blowing in the foam, and closing the holes.

Usually, all types of insulation are installed by the crews of insulation contractors, although rigid planks and batts may be placed by carpenters or common labor.

10.9. Intoxication

Most contractors, for safety reasons, have a policy of allowing no intoxicated person on the project site. For some it is a basis for discharge; others send the employee home to sober up. It is a well known fact that an intoxicated person on the site can cause injury to himself and/or others. It is now the position of some that intoxication is not a condition into which a worker willfully enters, but rather is an illness which he cannot control.

If one of your crew reports for work intoxicated, you must follow your employer's policy. It is recommended that he be discharged, or sent home, not for being drunk, but because he is a possible hazard to the progress of the work.

10.10. Inverts

The invert is the inside bottom of a pipe or the bottom of a flume or trough. It is the point at which the grade elevation is set. (See Figure 2.4.)

10.11. Inverted Siphons

Figure 10.1 will show you what an inverted siphon is and how it is used.

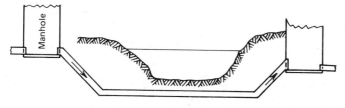

FIGURE 10.1. Inverted siphon.

10.12. Iron Pipe (IP)

The initials IP appear on maps to indicate property lines. They indicate the corners and line of your property and are legal markers set by surveyors. You must take care not to disturb these markers, and if your work is such that a marker must be disturbed, the surveyors should be advised in time to take references so that the marker can be replaced or moved to another location after the work is completed. The replacement of markers can be expensive, depending on how much surveying has to be done.

10.13. Ironworkers

Ironworkers erect the structural steel and place reinforcing bars, metal railing, and stairs. If there is a considerable amount of such work, they will have their own foreman. If there is a large quantity of structural steel, such as in a multistory building or a long-span bridge, the workers are quite likely to be Mohawk Indians as these men seem to have a unique sense of balance for working in high places.

11

11.1. Jackhammers

Jackhammers are used to break up concrete slabs, to cut through paving to allow trenching or other excavating, and to loosen shale, hard-packed soil, and rock. They are used to drive sheeting and poling boards in tunneling. They are usually powered by compressed air, but some small units are electric. They use bull-points, chisels, and spades, depending on the material being broken up.

There should be an oil injector in the hose line between the compressor and the jackhammer, and it should be kept full of oil. When operated in dusty areas or in cold weather, the exhaust ports will become gummed up unless care is taken to keep them clean. A stiff brush dipped in kerosene can keep them clean. If the internal valves stick, an easy way to clear them is to detach the hose and pour about ½ cup kerosene into the unit, connect it to hose, and operate the hammer in short bursts a few times.

11.2. Jacking

Jacking is one method of pushing a pipe through the earth to provide water, gas, or other service. It may be a casing, through which the service line will later be installed.

Jacking is also a method of raising a structure by use of a series of manual or powered jacks.

Jacking through the earth requires a special jacking machine, which is set into a trench at the proper line and grade. It will be supported on heavy timber to keep the pipe on line as it proceeds. The machine will require a solid backing to take the thrust as it pushes the pipe forward. When casing is used, some machines will rotate the casing as it is pushed forward. (See Section 3.25.)

When small pipe, 3 in. or under, is jacked, a blunt cap will be on the end to keep earth out. Larger casing pipe will have an auger inside to bring the loosened earth out for disposal.

When large pipe, 48 in. or more in diameter, is jacked, the earth will be hand excavated and removed by wheelbarrow or wheeled cart. The excavation will be slightly larger than the outside diameter of the pipe to lower the friction.

One of the biggest problems when jacking a pipe or casing under 3 ft in diameter is when large boulders or hard obstacles are encountered. Usually, the line will have to be abandoned and a new line will have to be attempted. In larger pipes, workers can get inside and remove the obstacle by jackhammer or blasting.

It is important that the machine and the pipe or casing be properly aligned before jacking starts. The longer the pipe section, the better. The timbers must be solidly in place so that they cannot move, and it is recommended that they be greased.

11.3. Jetting

When piling is being driven in tight soil, water is often jetted alongside to reduce the friction and allow the pile to be driven deeper. The riser pipes in a well-point system are driven by jetting. Water is sometimes jetted into a deep fill to increase settlement. Chemicals are jetted into the earth. (See Section 8.6.) Mud or grout is jetted under concrete slabs which have settled to raise them to the original grade. (See Sections 19.20 and 17.15.)

When jetting with water, there is usually is an excess of water that flows to the surface. This water will bear mud and silt, which be messy around the area, but can also cause problems if it flows into a stream containing fish. When possible, it is best to pump the excess water to an acceptable disposal area or to make a stilling area where the mud can settle.

11.4. Jitterbugs

The jitterbug is a tool made of small-angle frames formed into a rectangle about 10 in. × 36 in. which is covered with a sheet of heavy expanded metal and has a pipe or rod handle.

The unit is raised about 3 or 4 in. above a freshly poured concrete slab and slammed down so as to penetrate the concrete by about ½ in. and force the coarse aggregate down away from the surface. This action will leave only mortar for the top ½ in. of the slab and make the work of finishing much easier. Coarse aggregate at or very near the surface will leave only a very thin layer of mortar, which will dry out quicker that the thicker mortar and result in hairline cracks. The unit should be hosed clean after each use.

11.5. Job Organization

Any job must be well organized to operate effectively. As foreman, you must do the same with your crew. Get your instructions as far in advance as possible so you will have time to ask questions if necessary. Know what materials and equipment you will have to work with, and have everything on hand as needed. Arrange to have your equipment serviced at regular intervals and at times best suited to your progress. Know each worker's capabilities and where best to use him. Keep complete, accurate records on what was done and what materials and equipment were used. (See Section 19.3.)

11.6. Joints (See Sections 4.39 and 6.18.)

11.7. Joint Ventures

On very large and complicated projects that will require considerable financing and expertise, two or more contractors will work together in a joint venture. One person will be designated to head up the organization, and the workers of all contractors will work together as a unit.

11.8. Joists

Joists may be of wood, steel, or concrete. They are the small beams that carry the floor or ceiling. Joists may rest on beams or a wall. Most will have some kind of bracing between them, called bridging, which keeps the joist vertical when it is under load. (See Figure 3.3.)

11.9. Jumbos

A jumbo is a platform or movable scaffolding used at the heading of a large tunneling job to make a working platform for the drilling machines. There will be two or more levels, depending on the height of the heading. The machines will drill holes for blasting, and when the holes are drilled to the proper depth and the blasting powder is in place, the jumbo will be moved back on rails to a safe distance. After the blast has occurred and the loosened rock has been removed, the jumbo will be moved back to the new heading and drilling will be resumed. This operation is repeated as the tunneling proceeds. The jumbo must be sturdy enough to withstand the vibration of several drilling machines and must be able to move quickly.

11.10. Junction Boxes

There will be a junction box at each intersection of the metal conduit that carries electric power lines. After the conduit and boxes have been installed, the wire or cable is pulled through and connected at the boxes.

11.11. Jute

Jute is used to caulk the joints of tight sheeting in areas of high groundwater. A thin braid of jute is driven into the crack between the sheeting boards and rammed tight with a dull chisel or caulking iron. A sheet of jute is spread over freshly placed topsoil to prevent erosion (see Section 6.11.) Sometimes the sheets of jute are impregnated with a wax containing grass seed and fertilizer. Rain dissolves the wax, and the jute will rot away in a few weeks, leaving the seed to germinate.

12

12.1. Keel

The large (about ⅜ in. diameter) crayon used in construction is called keel. It is used to mark lumber, and surveyors use it to mark layout points and to put information on control stakes. The usual colors are blue, yellow, and red.

12.2. Keene's Cement

Keene's cement is a finely ground white powder which when mixed with water produces a smooth white mortar used to produce hard plaster finish. The mortar is also used to patch cracks and holes in plaster walls. When sealed, the smooth hard surface makes an excellent base for paint around sinks and washtubs. Like any mortar, it should be kept from drying out too fast or from freezing. When used as plaster, it will be applied with a rectangular steel trowel. For cracks or holes, a small pointed trowel or putty knife is used. The mortar should be forced deep into the crack and allowed to bulge slightly above the surface. After setting, the crack can be sanded smooth with fine-grade sandpaper.

12.3. Keys (Construction)

Keys as used in construction are projections that increase the friction between two surfaces. (See Figure 12.1.) The plans will show the type of key and

215

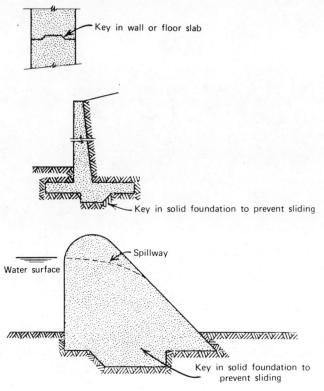

Key in wall or floor slab

Key in solid foundation to prevent sliding

Spillway

Water surface

Key in solid foundation to
prevent sliding

FIGURE 12.1. Keys in concrete construction to hold alignment and prevent sliding.

where they are to be placed. Before the concrete for the key is poured, the area should be cleaned of all debris, dust, or mud. The concrete should be plastic and well vibrated, as a strong, dense concrete is required to resist the shearing forces.

12.4. Keys (Locks)

On civil engineering projects during construction, there will be keys to tool sheds, fuel containers, fenced-in storage areas, and equipment. Any such keys entrusted to you should be kept safe on a strong key ring and attached to your belt with a strap. As soon as you no longer need a key, it should be returned to your superior.

The keys to all parts of the finished structure should be identified and turned over to the owner on completion of the project.

12.5. Kick Plates

A kick plate is a metal plate attached to the bottom 10 or 12 in. of a door for people to kick against when opening the door with their foot. Without the plate, the paint or varnish would soon be damaged.

12.6. Knots

Rope is used extensively on civil engineering projects as a temporary restraint and to raise and lower objects. It is important that the knots do not slip and cause accidents. It is recommended that you purchase one of the many books on the subject or read up on the subject in a local library. Using the correct knot for a particular purpose can make it easier to tie and untie the knot.

13

13.1. Labels

Some materials and permanent equipment will have labels attached certifying quality, model, and efficiency and giving instructions on operation and lubrication. These should be brought to the attention of the inspector or engineer so they can be preserved for the owner.

13.2. Labor Relations

(See Chapter 1, *Basic Ideas*.)

On most civil engineering projects, labor and management cooperate rather well so that the work gets done in a manner acceptable to the engineer and at a cost within the contractor's estimate. Good relations requires that the contractor pay proper wages and that labor give a proper day's work. Unfortunately, there are projects on which the contractor demands too much from labor and/or labor gives low production or demands compensation the contractor cannot afford to pay. If you are so unfortunate as to be on such a project, it is recommended that, although you are part of management, you try to be neutral if possible. (See Section 22.4.) As a foreman, you must see that your crew does work of acceptable quality within a reasonable time. You should treat your workers with respect and demand the same from them.

13.3. Ladders

Ladders used in construction should be inspected carefully at the start of each day, and if found defective, replaced. OSHA has very strict rules on

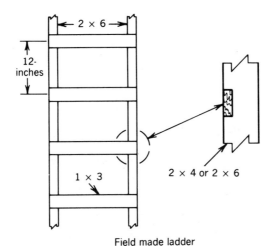

12-inches

2 × 6

1 × 3

2 × 4 or 2 × 6

Field made ladder

FIGURE 13.1. Field-made ladder.

the use of ladders and can levy fines for misuse. (See Section 16.2.) Defective ladders have been the cause of some serious accidents that could have been prevented by proper inspection. A perfectly good ladder can cause accidents if not properly supported at the top and bottom. (See Figure 13.1.)

13.4. Laitance

Laitance forms on the surface of freshly poured concrete when the internal water is forced to the top as the mass settles. The water will bring impurities and cement particles to the surface that, when set, will produce a hard shiny crust to which nothing will bond. The specifications will require that all laitance be removed from the surface before any additional concrete is poured on top of it. It is easily removed just before it sets by use of a rough brush or a wad of burlap. After setting, it will require a wire brush, chiseling, or sandblasting, depending on how hard it is, to properly remove it.

13.5. Lamping (Sewers)

Lamping to check the line and grade of a sewer line is the responsibility of the inspection team; however, some contractors lamp newly placed sections to assure themselves that the work is being installed correctly. The lamping is done by placing a strong light at a manhole and observing through the pipe toward the manhole. If the pipe has been laid to true line and grade,

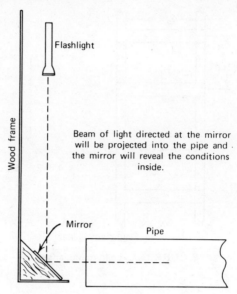

FIGURE 13.2. One method for looking into inaccessible areas. A beam of light directed at the mirror is projected into the pipe, or other area, and the mirror reveals the conditions inside.

the light will show up like a full moon. If the pipe is off line, you will observe a partial moon, and the less moon, the more the pipe is off line. By pouring several gallons of water down the pipe, any ponding will tell you if the pipe is off grade and about how much. If the pond is wide at the center and narrow at each end, the pipe will be low at that point. If the pond is wide at your end and tapers down to a point away from you, it means there is a high point in the pipe. This is true, provided the pipe is laid down the grade as is the usual case.

   If you are involved in laying sewer pipe, it is recommended that you check it, as noted, every few feet. Most contractors now use laser beams to set the sewer pipe, so that each pipe length is checked as it is laid. (See Section 13.7.) (See Figure 13.2.)

13.6. Landscaping

Landscaping is the art of improving the appearance of the surface of the earth by rearranging the contours, placing topsoil and grass seed, and installing trees, bushes, hedges, flower beds, and other items to beautify an area.

13.7. Laser Beams

The laser beam is a useful tool for setting and checking elevations on a construction project. It is also useful for laying sewer pipe, as it gives line and grade at the same time. Setting up and maintaining the beam on line and grade requires a trained crew. The beam is affected by changes in temperature and vibration. Once the equipment has been properly set and secured in position, the pipe is laid by placing a plastic target in the downstream end of the pipe, and the pipe is maneuvered so that the red dot from the beam is at the center of the target. The pipe is secured by tamping earth along and under the section until it is firmly set, keeping the red dot always at the center of the target as the tamping proceeds. There is one danger from a laser beam. You must never look directly into the beam, as it will damage your eyesight and could even blind you. When the laser is in use, there should always be a mechanic on hand to make frequent checks.

13.8. Late Arrivals (Workers, Materials, Equipment)

As foreman, you must do your best to see that all the labor, materials, and equipment you will need for the day's work will be on hand when required. Failure to have any of the three on hand when needed will slow the progress and increase production costs. If any of the three are missing, or if you believe they will be missing at the time they will be needed, you should confer with your superior at once.

The failure of one crew to complete its assignment on time could delay a whole section of the work. The foreman should not be held responsible for delays beyond his control.

13.9. Latent Underground Conditions

Latent underground conditions are likely to be encountered on any construction in a built-up area. The plans may show abandoned sewer, water, or utility lines with a note that the locations are only approximate. The contractor has to assume they are within a reasonable distance of where they are shown. If during excavation the lines are found to be much closer or farther away, so as to interrupt the operations of excavating, the contractor may have cause to file a claim. The specifications will state how such conditions will be handled. If your crew uncovers what appears to be an abandoned line, whether it shows up on the plans or not, do not damage it, but inform you superior and get instructions. Some such lines may appear to be abandoned, but may be used on occasion and, if blocked, backup could occur.

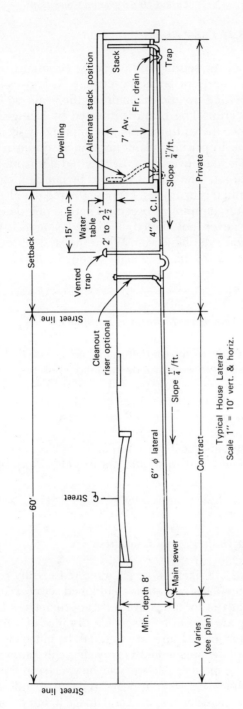

Typical House Lateral
Scale 1" = 10' vert. & horiz.

Distance Stack to Sewer	Depth of Invert of Street Sewer Below Water Table				
	6"φ	12"φ	15"φ	18"φ	22"φ
50'	9.0	9.4	9.7	10.0	10.5
60'	9.3	9.6	9.9	10.2	10.7
70'	9.5	9.8	10.1	10.4	10.9
80'	9.8	10.1	10.3	10.6	11.1
90'	10.0	10.3	10.5	10.8	11.3
100'	10.2	10.5	10.7	11.0	11.5

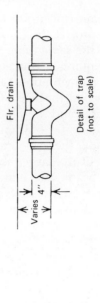

FIGURE 13.3. Typical house lateral.

If such lines are located in a position sufficiently different from that on the plans to warrant a possible claim, you must take measurements and data that can be used when presenting the claim. Photos are always good as records.

13.10. Laterals

Laterals are the sewer, water, or gas lines that connect the lines within a structure with the mains in the street. Water and gas lines are under pressure, so the grade is not important. The sewer flows by gravity, so it must have a proper grade from the structure to the main. (See Figure 13.3.)

13.11. Laying Schedules

Many water or sewer lines of large diameter are laid according to a schedule, which is a printed form giving the exact position and elevation of each section of pipe. Each section will be marked with an identifying number. The schedule will give the elevation of the invert at both ends of the section and the angle to which it is to be laid. Some sections will be angled or beveled to provide a change in direction. Each section must be laid in the sequence shown. (See Figures 13.4 and 13.5.)

PIPE LAYING SCHEDULE

BEGIN AT STATION 0⁺00 EXISTING 30″ MJ BELL LAY NORTHERLY.

	AVG LL	STATION	INV. ELV.
		0^+00	716.0
1 30″ MJ SPIGOT X 20″ FLANGE CONCENTRIC REDUCER	5.79		
1 VALVE (BY OTHERS)	0.39	$0^+05.79$	
1 30″ LJ BELL X 20″ FLANGE-REDUCER	5.37	$0^+06.18$	
1 STRAIGHT w/2″ OUTLET ON TOP C/L 12.00 FROM FACE OF BELL OPEN JT 3/8″ ON BOTTOM	16.02	$0^+11.55$	716.0
5 STRAIGHTS OPEN 1ST JT 1/4″ ON BOTTOM	80.10	$0^+27.57$	716.2
1 STRAIGHT w/16″ MECH JT BELL OUTLET ON RIGHT 4.00 FROM FACE OF BELL	16.02	$1^+07.67$	
2 STRAIGHTS	32.04	$1^+23.69$	
1 STRAIGHT OPEN JT 1/2″ ON LEFT	16.02	$1^+55.73$	
1 2-1/2″ BEVEL RIGHT OPEN JT 1/2″ ON SHORT SIDE	15.92	$1^+71.75$	
1 2-1/2″ BEVEL RIGHT w/2″ OUTLET ON TOP 4.00′ FROM FACE OF BELL, OPEN JT 3/8″ ON SHORT SIDE ROTATE CLOCKWISE FOR 3 – 39′ HORIZ DEFL RIGHT AND 0 – 12′ VERT DEFL DOWN	15.92	$1^+87.67$	719.1
1 2-1/2″ BEVEL RIGHT OPEN JT 1/4″ ON SHORT SIDE ROTATE CLOCKWISE FOR 3 – 39′ HORIZ DEFL RIGHT AND 1 – 05′ VERT DEFL DOWN	15.92	$2^+03.59$	719.3
1 2-1/2″ BEVEL DOWN OPEN JT 1/2″ ON SHORT SIDE	15.92	$2^+19.51$	
1 2-1/2″ BEVEL RIGHT OPEN JT 1/2″ ON SHORT SIDE	15.92	$2^+35.43$	
1 STRAIGHT OPEN JT 3/8″ ON LEFT	16.02	$2^+51.45$	
		$2^+67.37$	719.3

ALL LAYING INSTRUCTIONS GIVEN LOOKING IN THE DIRECTION OF LAYING, BELLS AHEAD. THIS SCHEDULE SUPPLIED ONLY AS A GUIDE FOR DISTRIBUTING AND INSTALLING PIPE AND FITTINGS. ADJUSTMENTS TO MEET FIELD CONDITIONS MUST BE MADE AS REQUIRED.
"MJ" = MECHANICAL JOINT   "LL" = LAYING LENGTH
"LJ" = LUGGED JOINT

FIGURE 13.4. Copy of a sheet of an actual laying schedule.

FIGURE 13.5. A reinforced concrete pipe water main being laid according to a laying schedule. Note surveyor with level and level rod with which to check each length of pipe as it is installed. On a vertical (or horizontal) curve, the ends of each section of pipe are beveled to make a tight joint all around. Note that the workers in the foreground are placing fine material as a bedding for the pipe. The three workers in the trench beside the last section of pipe are pulling fine material from the sides of the trench and placing and tamping it along the sides of the pipe.

The crane lifts the pipe by the cable sling, lowers it into the trench, and steadies it while the workers pull the joint tight by the come-along.

When installed, the pipe will be covered with about 2 ft of earth. This is enough to prevent freezing in a large main in which water is constantly moving.

13.12. Layout of the Work

If you are to be involved in the layout of work it is suggested that you obtain a copy of the book *Construction Measurements* by B. Austin Barry, that is a part of this series.

In laying out work, you must know where the control stakes are and how they relate to both horizontal and vertical measurements. You will need a standard 6 ft folding carpenter's rule and/or a steel tape. (Cloth tapes are not recommended.) If you work with the surveyors, you will note that their steel tapes will show feet and tenths of feet. You must learn to transpose from tenths to inches (1 in. equals 0.08333 ft). On some layouts you will need a plumb bob, which is easy to use once you learn. The tape or rule must be kept level, and the plumb bob must be still as you mark the point. It is handy to learn to use a builder's level for vertical layout. Figure 13.6 will show you how to use a rule or measuring tape. All layout should be

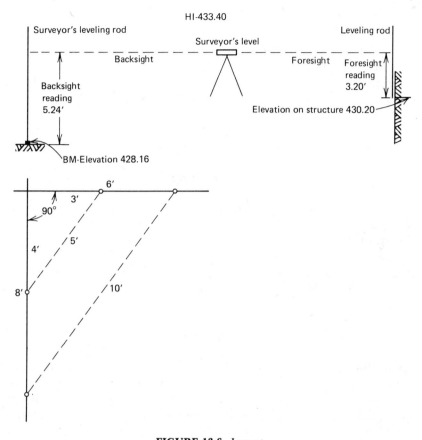

FIGURE 13.6. Layout.

done with care, and an immediate check is recommended since an error could cause much work to be torn down.

The layout should be sufficient for a full day's work, but not so far ahead that it might be disturbed before it is used. The process of establishing an elevation is simple addition and subtraction. For example, referring to Figure 13.6, the surveyor's level is held on the bench mark (BM), and the distance above the BM is read on the leveling rod. This distance added to the elevation of the BM gives the height of instrument (HI). Thus, 5.24 added to 428.16 gives 433.40, which is the HI. If the elevation on the construction is 430.20, the surveyor must read 3.20 on the rod (433.40 − 430.20 = 3.20). As the surveyor reads 3.20 on the rod, a mark is made at the bottom of the rod to give the desired elevation of 430.20. If you run through the preceding exercise a few times, you will quickly understand what to do.

The corners of a structure are usually right angles; they must be exactly 90 deg if the various parts are to fit properly. A simple method of checking for a right angle is to use a measuring tape or leveling rod and mark out 6 ft on one side and 8 ft on the other. The distance between the two points (hypotenuse) will be 10 ft if the angle is exactly 90 deg. In close quarters, 3, 4, and 5 ft, respectively, may be used.

13.13. Leads (See Section 17.15.)

13.14. Leaks

Leaks in a structure during construction or during the one-year guarantee period can be very costly for the contractor. Leaks through roofing and walls, around doors and windows, and at points where piping goes through a wall can result in damage to the interior wall finish and ever to furnishings. Leaks can discolor ceiling and wall plaster and may cause it to disintegrate.
For further information, see Section 10.3.

13.15. Liability

Liability for injury or property damage during construction is with the contractor. He will carry worker's compensation and other insurance to cover such problems. This is one of the costs of doing business. However, the fact that there is insurance coverage should not be a reason for laxity in safeguarding against accidents. A history of frequent accidents will raise the insurance rates and increase OSHA's surveillance of the operations. You should continually study your crew's methods of operation in an effort to anticipate problems and make corrections to avoid them. If your crew is accident prone, it will affect your rating with your superiors.

13.16. Lightweight Aggregate

Lightweight aggregate such as cinders and slag is used to make lightweight concrete and as insulation in walls and on roofs. Many precast concrete roof panels are made of lightweight aggregate, as the loads are light and the units are easy to handle.

13.17. Lightweight Concrete

Lightweight concrete weighs from 40 to 75 lb/ft^3, while regular concrete will weigh 140 lb or more. Lightweight concrete does not bond as well to reinforcing steel as regular concrete does, so it will have less strength. It is used as fill, as insulation, and to provide a drainage slope on flat roofs.

13.18. Lining (Aggregate)

Aggregate classified as lining is sand or well graded run-of-bank gravel used under and around sewer and water pipes on unstable or rocky subgrades (see Section 20.61). The inspector or engineer will specify the depth under the pipe, which will run from a few inches to a foot or more depending on conditions. All lining should be placed in well compacted layers under and up around the sides of the pipe. The lining is usually carried from the stockpile to the trench by front-end loaders and should be dropped alongside the pipe and not on the trench bank where much would be wasted.

Lining is also used under concrete foundations and basement floor slabs. Under muddy conditions, the lining will provide a dry working surface and keep the reinforcing bars clean. Lining under concrete slabs must be compacted to avoid settlement.

Surface

Surface lining is used to prevent leaks and erosion in reservoirs, canals, storage ponds, and such. In past times, and occasionally now, clay was used for the lining, but in recent years, Portland cement or bituminous concrete is most often used. Plastic sheets, which will have the seams cemented together to prevent leaks, are also used. Either kind of concrete lining can be placed by crews experienced with that kind of material. The plastic sheets are placed by specially trained crews. The biggest problem when placing such lining is waterproofing the points where pipes or structures pass through the lining. A prime coat should be painted on the pipe or structure to provide a better bond. If it is necessary to walk on the sheets, planks should be placed across them to prevent puncture.

Old reservoirs with concrete lining may have the surface restored by applying Gunite (see Section 8.21).

13.19. Lintels

Lintels may be made of wood, steel, or reinforced concrete. They should be well founded on their bearing surface. They are used over openings for windows, doors, and louvers and over openings in walls within structures. Weak or poorly placed lintels can result in unsightly cracks over the opening. (See Figure 13.7.)

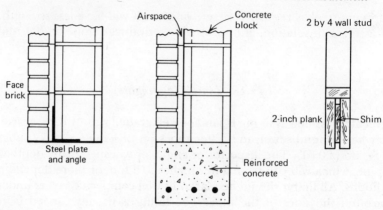

FIGURE 13.7. Lintels. The size of a lintel depends on the span and loads to be supported.

13.20. Liquidated Damages

Most contract documents will contain a clause specifying a penalty if the contractor does not complete the work on time. It will not be called a penalty, as most states forbid the use of that word under such conditions. The cost will be labeled "liquidated damages." Regardless of the wording, if the work is not completed on time, and there are no extenuating circumstances, the contractor may have to pay a specified sum for each day of overrun. That is one of the reasons why you should keep the work of your crew on schedule and not delay your own or another crew's progress.

13.21. Lock-Joint Pipe

Lock-joint pipe is a type of heavy reinforced concrete sewer or water pipe used where high pressures are expected. The bell and the spigot ends will have metal inserts with grooves for heavy gaskets. When installing this type of pipe, the gasket should be inserted in the groove, and then the whole metal surface and the gasket should be well lubricated.

13.22. Loss Abatement

General contracting on civil engineering projects is a very competitive business and sometimes requires a very small profit margin to make the bid low enough to get the job. There are always the problems of bad weather, late delivery of materials, and breakdown of equipment, none of which can be anticipated accurately. To compensate for these uncontrollable losses, most contractors will have a loss abatement program where he will hire highly skilled supervisory personnel and workers and conduct regular training sessions to acquaint them with his methods of operation and to plan ways to lower the costs of production.

As a good foreman, you should always be studying each operation, searching for a way to increase production without overburdening your workers. You should also look for ways to increase the efficiency of your equipment and the handling of materials.

A foreman who keeps coming up with cost-saving, but practical, ideas will have no problem with promotions.

13.23. Louvers

Louvers are ventilating units set in a wall the same as a window or door, except that the louver has slats set a few inches apart and sloped to keep the rain out. Most will have a screen on the outside to keep birds, insects, or such things as pieces of paper from getting inside. Some louvers will ventilate in the same way as a window, while others admit air for air-conditioning units or pressure ventilation. When used with large air-conditioning systems, a large volume of air will be moved at high speed, so the louver frame must be securely anchored, and the crack between the frame and the wall must be well caulked.

13.24. Lubrication

Proper lubrication is a must, whether it is for the contractor's own equipment or for units to be installed as part of the contract. The contractor will usually arrange to have his own units serviced at regular intervals. (See Sections 6.9 and 6.10.)

Units that are to be installed as part of the contract should come properly lubricated and usually with a supply of the proper lubricants. However, sometimes a piece of equipment will come through with little or no servicing. Each unit should be checked before starting it up to ensure that all places are either greased or oiled as required. It is best to turn the unit over by hand to see that it is free and ready to operate. Many pieces of equipment will come with instructions on how to service them, and these instructions should be preserved and turned over to the owner.

13.25. Lugged Pipe

Sewer and water mains that are to be under heavy pressure, particularly those of large diameter, may be lugged instead of having thrust blocks at changes in alignment. The lugs will be at each connection between the sections and set so they can be bolted together.

13.26. Lump Sums

Some smaller projects may be undertaken for a lump sum. That is, the work will all be done for a single specified sum of money.

14

14.1. Maintenance

(Contractor's Equipment)

The proper and timely maintenance of the contractor's equipment can ensure maximum production from the units. Most contractors will have a mechanic responsible for routine servicing and fueling. Many projects refuel during the lunch period, when the equipment is idle. Unusual sounds and/ or difficulty in operating a machine can foretell a breakdown. You and the machine operator should be alert for problems and advise your superiors. It is better to replace a machine before it breaks down and requires a major overhaul. Things that can indicate future problems are; oil or grease leaks, unusual noises in the motor or gear boxes, hot or chattering bearings, backfiring of the engine, signs of wear or breaking of some strands of a cable on a hoist, and signs of wear or squeaking of the brake shoes on the hoist drums. All cutting tools should be kept sharp as they not only cut better, but are less likely to slip.

Permanent Equipment

Permanent equipment installed as part of the project may require maintenance by the contractor during the one-year guarantee period, at the end of which it must be in proper working order. Proper installation, such as proper alignment of parts, proper lubrication, equal tightening of anchor bolts, and full bedding of the machine base on its foundation, can reduce the maintenance requirements.

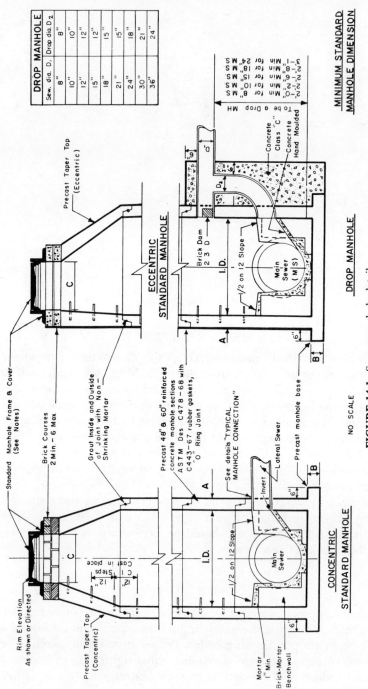

DROP MANHOLE	
Sew. dia. D₁	Drop dia. D₂
8"	8"
10"	10"
12"	12"
15"	12"
18"	15"
21"	15"
24"	18"
30"	21"
36"	24"

MINIMUM STANDARD
MANHOLE DIMENSION

2'-0" Min for 8" M.S.
2'-6" Min for 10" M.S.
2'-6" Min for 15" M.S.
2'-8" Min for 18" M.S.
3'-1" Min for 24" M.S.

ECCENTRIC
STANDARD MANHOLE

DROP MANHOLE

CONCENTRIC
STANDARD MANHOLE

NO SCALE

Precast Taper Top (Eccentric)

Precast Taper Top (Concentric)

Standard Manhole Frame & Cover (See Notes)

Brick Courses 2 Min – 6 Max

Grout Inside and Outside of Joint with Non-Shrinking Mortar

Precast 48" & 60" reinforced concrete manhole sections ASTM Des C478-68 with C443-67 rubber gaskets, "O" Ring Joint

See details "TYPICAL MANHOLE CONNECTION"

Precast manhole base

Rim Elevation As shown or Directed

Cast in place

Steps 12" 12"

Mortar 1" Min

Brick-Mortar Benchwall

Lateral Sewer

Invert

Main Sewer

Brick Dam 2-3 D

Concrete Class "C"

Concrete Hand Moulded

Main Sewer (M.S)

To be a Drop MH

½ on 12 Slope

I.D.

FIGURE 14.1. Sewer manhole details.

232

14.2. Manholes

Manholes are placed at all changes in line and/or grade in sewer lines and at the junction points of underground cables or conduits. (See Figures 14.1–14.3.) Manholes are made of precast or cast-in-place concrete, concrete blocks, brick, and sometimes metal or plastic. They provide access for maintenance and repairs.

The manholes built of blocks or brick will be constructed after a concrete base has been poured and the pipe has been laid. The units will be set around the pipe and the void will be filled with mortar. The cast-in-place concrete may be poured around the pipe, or a hole may be left and the pipe will be installed later and mortared in place. The other prebuilt manholes will have holes left for the pipe, which will be sealed in place by mortar and brick.

FIGURE 14.2. Precast sections used as either manholes or catch basins. The base section is set on a firm, subgrade or a concrete slab, and intermediate sections are added to provide the height required. The opening in the wall of the base section will depend on the size of sewer pipe used. In this case, it is 24 in. The opening outside of the pipe will be filled in with brick and mortar. In this case, the units were used as catch basins, so the flat covers shown were placed on top and a cast-iron grate was set in the rectangular hole. If the cover was a little too low, a brick wall would be placed around the rectangular hole, and the grate would be set on top of that. The same type of unit is used for manholes, but with the top section as shown in Figure 14.1.

FIGURE 14.3. A close-up of the base section where the manhole steps can be seen through the opening.

When backfilling around a manhole, care must be taken to bring up the fill evenly on all sides to keep it from tipping out of plumb. When the manhole is in a traveled way, the manhole frame and cover must be set slightly, about ½ in. below the surface to prevent it from being hit by a snowplow in cold regions. When the frame and cover are set too high or too low, they cause bumps for traffic.

When the grade of the connecting pipelines varies by over a foot, it is customary to install a "drop" manhole. Figure 14.1 indicates the design.

Manholes for service lines will have conduits connecting them, through which the lines can be installed and replaced when necessary. Such manholes usually have a sump pit for pumping out water which seeps in during rainstorms.

14.3. Masonry

Masonry includes brick, concrete blocks, tile, and stone. The installation of any of these items will be done by masons, who will have their own foreman. Laborers will be used to handle materials and assist in erecting and moving scaffolding. They will clean up the excess mortar that falls on the ground and scaffolding planks and is left on the mortar boards as work progresses. They may do the cleaning of the surface after the work is done. The cleaning should be done before the mortar can set, for once it sets it is almost impossible to clean it off the rough surface of the brick or blocks. The cleaning will be done with a stiff or wire brush and water, to which a 10% muriatic acid solution has been added. The solution should be applied quickly and immediately hosed off with clean water. If all the mortar has not been removed, the operation should be repeated. The solution should not be left on the surface for over a few minutes.

In hot weather, the masonry units may have to be dampened, but not wet with free water. This is best done by using a fine spray of water and then covering the units with damp burlap. On many projects, the joints of mortar between the units will require "tooling," which means the joint is smoothed out by a tooling iron into one of four patterns. (See Figure 14.4.)

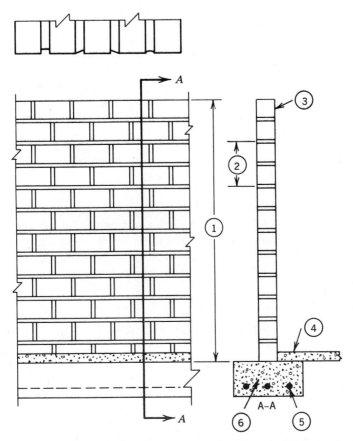

FIGURE 14.4. (*a*) The four usual ways of *pointing up* the joints. (*b*) A concrete block or tile wall.

If face brick is being placed over a block or concrete wall, weep holes to allow condensation to leak out will be spaced 10–12 ft apart at the bottom of the wall. These must be cleaned of all mortar before it sets. All joints between the units, both horizontal and vertical, should be of a uniform thickness, which will be specified in the contract.

The mortar will be a mixture of mason sand, mason cement, and water, with the exact mix specified. The sand and cement will come in sacks in the correct mix, so only water must be added.

14.4. Materials

You will seldom be concerned with the quality and acceptability of the materials to be used in the project. The engineer or his inspector has that responsibility. You will be concerned that the materials on site are not damaged after arrival. Lumber can become warped if not stacked properly. Aggregate can become useless by being covered with dust, mud, and sometimes oil. Reinforcing bars are bent when run over by trucks or other vehicles. Equipment can become damaged if hit by moving vehicles or if it is stacked too high and topples. Some materials are paid for by the cubic yard or ton as delivered to the site. You should obtain the delivery slip for each load, check to see that what is on the ticket represents what is being delivered, and turn the tickets over to your superior.

14.5. Mats (Concrete)

Concrete mats are sometimes used under cradles for pipelines laid in unstable areas. (See Figures 14.5–14.8.) The mats may be reinforced with steel bars of mesh, and if so, the reinforcing will extend beyond the day's pour in order to connect it to the next pour. Such reinforcing should be in the proper position and clean before the next pour starts.

14.6. Mats (Construction)

Mats are on construction for two purposes. A mat of heavy timber is used to support equipment when used in unstable areas. (See Figure 14.9.) Mats of woven cable or rope are used to cover an area during blasting to prevent pieces from flying out and injuring personnel or damaging property. The timber mats will be made of 10 × 10 in. or larger timbers, with several bolted together as shown in the figure. These mats will usually have a cable at one end for ease in hoisting and will be placed by crane or backhoe.

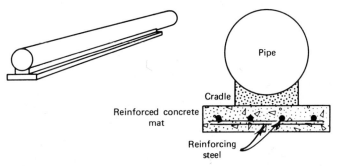

FIGURE 14.5. Concrete mat. This drawing shows an installation similar to that pictured in Figure 14.6.

FIGURE 14.6. Concrete mat. One method of installing a sanitary sewer in an area of unstable ground. The pipe is about 12 ft below ground surface. There is 2 ft of compacted run-of-bank gravel, a 9-in.-thick reinforced concrete mat, and a concrete cradle for an 8 in. A/C sewer pipe.

FIGURE 14.7. Steel bars set in place for a reinforced concrete mat to support a 24-in. sewer pipe.

FIGURE 14.8. Some of the problems encountered when installing a trunk sewer through a swampy area.

Note the well-points on the right-hand side. As the banks of the trench are bleeding water, extra pumps are needed to remove the accumulation from the subgrade before the concrete mat can be poured.

Note the mason's line and batter boards over the trench for setting line and grade of the pipe.

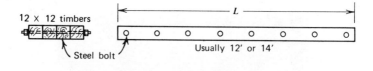

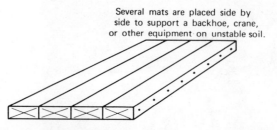

FIGURE 14.9. Mat construction. To make a construction mat, 12 × 12 timbers are bolted together as shown. Several mats are laid side-by-side to make the width required for the machine that is to be used.

14.7. Mats (Mud)

Mud mats are quite often required where heavy foundations require reinforced concrete. The mat will be a 4–6-in. unreinforced concrete slab placed on the subgrade to prevent mud from forming and getting on the bars as the workers place them. The slab will also make a better surface for setting the chairs for holding the bars in place.

14.8. Measurements

(See Section 13.12.)

14.9. Merit System

Some contractors use a merit system to improve production and maintain good morale among the workers. A reward can take various forms. Some reward above-average production with an increase in salary, others with a year-end bonus, and still others with increased paid vacations. To have some kind of merit reward is good business. Even unskilled workers will respond to justified praise of work well done. Even the foreman can gain from his own little merit system by recognizing work well done and seeing that the supervisors are informed of it. Any praise of a worker should be given in the presence of all the crew.

14.10. Mesh (Wire)

Wire mesh comes in various gauges, from small to heavy. The light gauge will come in rolls several hundred feet in length. The heavier gauges will come in sheets 8–12 ft in width and 10–20 ft in length, with several sheets in a bundle. The small-gauge mesh is used for reinforcing in Gunite, stucco, and light concrete slabs. The heavy gauge is used as reinforcement in roadways and other heavy slabs, as security over windows and around machinery, and as protection over openings in floors. When used in continuous concrete slabs, the sheets should overlap at least two squares. (See Section 6.17.)

14.11. Metric System (See Appendix)

There has been a movement for some time to convert to the metric system, but the movement does not seem to be making much headway at the present time. If you are involved in a project that uses the metric measurements, it is suggested that you obtain one of the many charts and books on the subject now available in most bookstores.

14.12. Metric Tables (See Appendix)

14.13. Millwork

Millwork covers wooden items built and finished by machine at the mill, including doors and windows and their frames; stairways including the railings, stair treads, and landings; wainscoting; hardwood-finish flooring; kitchen and other cabinets; baseboards; and other built-in fixtures.

On civil engineering projects, many of these items will be used in quarters for maintenance crews, offices, laboratories and control rooms. The units may come finished or unfinished. If unfinished, the items should be kept in the original containers until ready to be installed. If the finish is to be paint, the units should be primed as soon as removed from the container; if varnished, they should be handled with gloves until stained and sealed, as oil from the fingers will prevent the stain from penetrating the wood. If the units come finished, they will probably be covered with paper, which should be left in place until they are ready to be installed.

14.14. Mistakes

Everyone makes a mistake occasionally, and each one costs the contractor something. Those that are corrected immediately usually cost much less to correct. Most of us do not like to admit we made a mistake. If you or a member of your crew makes a minor mistake that will not materially affect the project, correct it immediately without fanfare. However, a mistake that can affect or delay the project should be reported to your superiors at once, and a determination should be made on how best to proceed to make corrections. Never try to cover up a mistake; even a minor one that you corrected immediately should be reported, telling what was done and why.

You should be constantly on the alert to anticipate possible mistakes so as to keep them to a minimum. A contractor cannot afford to keep a foreman who makes numerous mistakes.

14.15. Mixers (Concrete)

On most projects today, the concrete will be supplied by a commercial supplier by use of Redi-mix trucks, although on very large jobs, like a concrete dam or bridge, the contractor may have his own mixing plant. In such a case, he will have a trained crew to operate the plant. On-site concrete may be delivered to the point of deposit by concrete pump, trucks, narrow-gage railway, or overhead bucket and cable. (See Section 4.35.) If you are in charge of a concrete mixing crew, you must see that the aggregate has been approved, be sure it is clean, and have some method of measuring the correct amount for each batch. The mixing time must be controlled. The mixer must be cleaned after each day's use.

14.16. Monuments (Property Markers)

Replacing damaged property markers is a costly undertaking unless the control points happen to be close by. If your crew is assigned to an area close to the property lines, you should check for and prominently mark any monuments. A monument may be a concrete post, an iron pipe, or even a wooden stake.

14.17. Mortar

Mortar used for laying brick, cement blocks, wall and floor tiles, and building stone is made of mason cement and mason sand with enough water to make a plastic mix. Mason sand is finer than concrete sand and mason cement sets more slowly than Portland cement. The mix will be "fatter" so that it will stick to the masonry units.

The mortar should be dry enough so that as a unit is placed the squeezed-out mortar will not run down the face, particularly if it is brick, as it is difficult to remove.

Mortar stains on the face of tile can be removed by scrub brush and water if done at once. If the stains have set, you might never remove them. Mortar stains on brick or blocks can be removed before they set by scrub brush and water with a 10% muriatic acid solution, flushing them off with clear water as soon as the area is scrubbed.

14.18. Motivation

A good foreman will motivate his crew to do quality work at a production rate of which they can be proud. He will do this by making the working conditions as good as possible, by recognizing superior work, and by commenting favorably on it to other members of the crew and to his superiors. No flowery speeches: just a quiet statement of the facts. He will be friendly with each member of the crew, but a close buddy of none. Almost everybody enjoys and will respond to a little praise now and then, and it should be given when justified and only when justified.

14.19. Mud (Disposal)

Mud can be a costly problem on a construction project, particularly in built-up areas. Muddy water discharged onto a traveled way can be slippery while wet and dusty after drying, either of which can cause traffic problems. Mud-carrying water can clog catch basins, or if it flows into a stream containing fish, it can kill them and is a violation of the law in most areas. If a structure is flooded, mud will settle on the floor and walls and anything in the area. An easy way to remove mud in a flooded structure is to place a mud pump in a low area of the structure and hose everything down with a garden hose. The stream of water will keep the mud in suspension as it is being pumped out. The hosedown should continue until the pump discharges clear water.

Wet earth hauled in trucks that are not watertight can leak muddy water onto the traveled way.

14.20. Mud Pumps

A diaphragm pump will be used to dewater an excavation when the water carries mud, silt, small stones, and other debris. The pump will have a 3

FIGURE 14.10. Pressure-type mud pump used to raise floor or pavement slabs or fill other voids. A mixture of clay, sand, and water is placed in the pug-mill-type mixer through the hopper on top. When the mix reaches the consistency of thick mud, it is forced under the slab by high-pressure piston pumps. The mixture flows through a special reinforced hose to a nozzle that clamps into the hole in the slab. A pattern of holes is drilled through the slab, and as each hole is filled, the pressure is shut off and the nozzle is moved to the next hole. By repeating this procedure several times, the slab can be slowly raised to the desired position.

or 4 in. suction and discharge hose. The suction hose is reinforced with steel wire to prevent collapse. The valve may be a rubber flap valve or a hard rubber ball. As the diaphragm is raised, it creates a vacuum, which opens the valve and draws the water in; then, as the diaphragm is lowered, it closes the suction valve and opens the discharge valve, forcing the water out through the discharge hose. Because the unit works by vacuum, the diaphragm must be tightly sealed and have no cracks or pinholes. The suction hose must be tightly connected to the pump and also have no cracks or pinholes. There must be a screen on the suction hose to keep out any object too large to pass through the valves. The openings in the screen will keep out the larger objects. Because the discharge will carry silt, sand, small stones, and other objects, care must be taken to direct the discharge to a point where the material can settle without causing problems. After a day's work, the pump, especially the valve seats, should be thoroughly cleaned.

Another type of mud pump has piston-type plungers, which force the mud out under considerable pressure. Such pumps are used to force mud or cement grout into voids under slabs and to raise roadway and floor slabs that have settled. (See Figure 14.10.)

14.21. Mulch

Most specifications will call for a mulch of straw or hay to be spread over a newly topsoiled and seeded area. A machine will grind up the hay or straw and blow it out onto the surface at the specified depth. As even a light wind can blow the mulch away, it should be dampened by a light spray of water. Heavy spraying will wash the mulch into piles. The mulch should be kept damp until the grass starts to sprout. (See Sections 6.11 and 11.11.)

15

15.1. Nails

Nails come in various types and sizes, depending on their use. Wire nails are used on wood framework. Small-headed finish nails are used on wood trim and the head is countersunk with a nail set. Hard-metal cut nails are used to attach wood or metal strips to concrete or masonry. Double-headed nails are used on wooden formwork for ease in pulling when dismantling. The nails may be hand driven or a power hammer may be used. When hand driven, the nail should be "set" first and then driven down; otherwise, it might fly off and hit another worker. (See Figure 15.1.)

15.2. New Workers

When a new worker is assigned to you, you may be advised of his qualifications and limitations. If not, you must find out for yourself so that he can be placed where he will be most efficient and fit in with the other workers. On occasion, the new worker may be harassed by the old members of the crew, sometimes in fun and sometimes because they resent a new member who might be better than they are. Keep close watch on a new worker for several days and arrange to put him where he fits in best with the others. You should not allow harassment. If you have organized a harmonious and efficient working crew and the new worker just does not fit in after a reasonable try, you should ask your superiors to move him to another assignment.

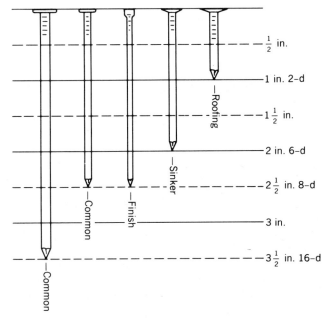

FIGURE 15.1. Nails most commonly used on civil engineering projects.

15.3. *Night Construction*

Night construction is always more hazardous and usually less efficient than in the daytime. Night work can result from a problem which slowed day work that must be continued until a section of the work is completed, or there may be a regular night shift. Some large projects will have three full shifts around the clock.

Even on a well lighted project, there will be shadow areas where a worker can step on or bump against an object that could cause injury. It is recommended that heavier shoes and trousers be worn for night work. There should be battery- or generator-operated emergency floodlights in case of a power failure in the area.

While most workers will produce as well at night as in the daytime, there will occasionally be one who, either because he just cannot adjust to night work or is just lazy, may step into the dark corners for a short nap. This not only affects the crew's production, but can be dangerous if equipment is operating in the area or someone decides to dump material on the spot. You should take an occasional headcount.

If you and your crew are on extended night work, you should each be more concerned about your physical well being, getting plenty of sleep and eating a well balanced diet. It is much more difficult to stay alert at night.

15.4. Noise

Civil engineering projects are noisy at best; however, a considerable amount can be eliminated with a little effort. If the project is in wide-open spaces, there is usually no problem, but if the work is in or near built-up areas, there can be considerable complaining.

Some of the noises that can be eliminated or reduced in volume come from unnecessary shouting, "gunning" the engine on trucks or other vehicles, backfiring out-of-tune engines, unnecessary blasting, hammering on structural steel, and sounding horns excessively. Not all noise can be eliminated, but much of it can be reduced, and an effort should be made, if only for public relations. You should not try to reduce noise to a point where it slows production. Many localities have laws that limit the volume of noise and forbid construction noise between certain hours.

15.5. Nonshrinking Mortar

Nonshrinking mortar is used under the bearing plates of steel columns and under the bases of machinery. Mason mortar shrinks slightly as it sets, so it could cause uneven settlement. It can be made nonshrinking by mixing in a chemical, of which there are several on the market. A nonshrinking mortar mix can be purchased dry and mixed with water as needed. It is recommended that rubber gloves be worn, as some of the chemicals can burn the skin. The mortar should be mixed in small batches and rammed in place quickly.

15.6. Nonskid Surfaces

A nonskid surface can be made on a concrete slab by brooming the surface just before the concrete sets, or the surface can be sprinkled with coarse, sharp sand just before it sets. A painted surface can be treated with a sprinkle of coarse sand before it sets. Steel stairs and other surfaces can be made of steel plates with small knobs made on them as they are hot rolled. Rubber mats with ribbed surfaces are used on stair treads, and abrasive metal strips are sometimes cast into concrete stair treads. Roadway surfaces are cut with carborundum saws to provide small grooves which make the curves nonskid.

15.7. Notes

It is a good habit to carry a small notebook and jot down a few notes on anything unusual that occurs on the job and to later transfer the notes to your diary. (See Section 5.9.)

15.8. Not to Scale (NTS)

Many construction plans will have the letters NTS beside detail sketches. This means the sketch is not drawn to scale, so you must rely on the printed dimensions and not try to use a scale or ruler.

16

16.1. Obstacles

Unanticipated obstacles can delay the progress of the work and may be the basis of a claim for extra work. (See Section 13.9.) Such obstacles may be reusable or saleable. If you run into such items, you should confer with your superior. The specifications usually indicate the disposition of such items as sewer and water pipe, power cable, cut stone from foundations, heavy timber used as sills, or anything that can be used on the project or sold.

16.2. Occupational Safety and Health Administration (OSHA)

The Occupational Safety and Health Administration (OSHA) is an agency of the federal government. Its inspectors visit construction projects to see that the health and safety regulations are being followed. Heavy fines can be levied for violations. The inspector can order all or part of a project shut down if, in his opinion, there is a violation, and work cannot be resumed until corrections are approved.

OSHA regulations have to do with all operations of a project and especially the use of machinery, excavating methods, the protection of excavations, scaffolding, and hoists. The inspector will check on the wearing of hard hats, the slope of trenches, warning signals on moving equipment, and the use of ladders.

Do not argue with an inspector if he comments on your operations. Listen to what he has to say and then confer with your superior if the inspector is not satisfied with the work.

Other items the inspector will be concerned with are loose boards laying on the ground, especially if they have nails in them, proper guard railings, worn hoist cables, proper sanitary conditions, and safe drinking water.

16.3. Odors

Odors can be a problem when the construction site is over or near an old sanitary landfill or where chemicals from industrial waste have leached into the soil. The odors can cause nausea, headaches, and watery eyes. Quite often, the odors are not detectable until the ground is excavated or piling is driven. If the odors are not too strong, it is sometimes possible to reduce them to a point where they are no longer offensive by opening the area and allowing it to sit for a few hours. If that does not work, it is best to engage a chemical engineer or chemist to analyze the odors and prescribe a chemical solution that, when sprayed over the area, will reduce or eliminate the odors.

Occasionally a worker, particularly an older, unskilled one, will neglect his personal hygiene to the point where it is offensive to those working near him. Quietly tell such a person to take a bath and wash his clothes before coming back to work.

16.4. Official Visitors

You should keep an accurate record of all official visitors who come to the site of your work and, particularly, note any conversations or instructions given to you. If you do not know, you should ask for the name and title of the visitor, you should record the time of day and how long he stayed on the site. If you were given instructions other than those already given to you by your immediate superior, you should advise that superior before proceeding.

There have been cases where problems on the job site have resulted in court action where higher-ups have tried to place the blame on the foreman. Many courts ruled that high officials had visited the site and observed, or had the opportunity to observe, conditions and so must assume all or most of the blame. Under such conditions, a clear, detailed record is invaluable.

16.5. Offset Stakes

Construction control stakes are offset from the true line so as not to be disturbed during operations. If you are to lay out your work from these stakes, you must know what each stake represents and how far you must measure both horizontally and vertically to establish a starting point for your work. (See Sections 2.10 and 13.12.)

16.6. Optimum Moisture Content

The optimum moisture content of the soil is that amount which will allow maximum compaction. Too much moisture will cause softness and mud; too little moisture will cause the soil to crumble and not compact properly. The optimum moisture is determined by laboratory tests taken at the site. If there is too much moisture, it is usually necessary to wait until it dries out. In some cases, lime is spread over the area to hasten the drying. If the soil is too dry, the area will be sprinkled with water. On highway construction, a water truck will sprinkle a path about 10 ft wide. On some projects, a garden hose is used to spray the area.

A simple way to judge the approximate condition is to take a handful of the soil and squeeze it tightly. If the soil is too wet, it will run out between your fingers and feel wet in your hand. If it is too dry, it will be difficult to compact it into a ball. If you can compress it into a ball that stays firm as you release it, it will be pretty close to optimum. (See Section 4.32 and Figure 6.3.)

When the job is too small for a tank truck, you can improvise with a sprinkler such as is shown in Section 6.6, Figure 6.3b.

16.7. Orangeburg Pipe

Orangeburg is a trade name of a bituminous fiber pipe used mostly for drain lines. Some lengths of the pipe will have perforations for about half of the circumference and will be used in sanitary filter beds. (See Figure 7.3.)

16.8. Orders (See Sections 4.20, 6.22., and 20.55.)

16.9. Or Equal

Many specifications specify materials or equipment by brand name, but allow other items to be used by inserting an "or equal" clause, meaning something else can be used if it is "equal" to the item specified. These decisions will be made by the engineer and given to your superiors.

16.10. Oxidation Pond (See Section 2.4.)

17

17.1. Painting

Painters will have their own foreman, who will have come up through the ranks and had experience in techniques such as surface preparation, color mixing, sealing, priming, and application.

In general, it is required that the surface be clean and smooth and that there be no dust, scales, or loose particles. The weather should be moderate, above 40° F and preferably not above 90° F, with no threat of rain in the near future.

Masonry surfaces should be damp but not wet to receive water-based paint, and after the paint has been applied, it should be kept damp for several hours.

Wood surfaces should have all cracks filled and all knots sealed.

All paint should come to the job site in the original containers and the manufacturer's directions should be followed.

17.2. Panic Bars

Panic bars will be found on almost all exit doors in public buildings and many industrial plants. On the inside of the door will be a metal bar that reaches the full width of the door and is connected to the door latch. Anyone pushing against the bar will unlatch the door and cause it to open. The latch on the outside will not allow anyone to enter without a key. In case of a panic, people crowding against the door will cause it to open, hence the name.

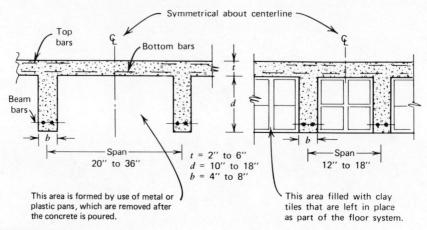

FIGURE 17.1. Pan concrete forms. Beams are formed by use of metal or plastic pans set as shown. The pans are removed after the concrete is poured and set. On some projects, clay tiles are used instead of the pans. The tiles are left in place to form a smooth ceiling. Either method forms T-beams made of reinforced concrete.

17.3. Pans

On civil engineering projects, "pans" refers to two things. The large earth scrapers are sometimes referred to as pans, and metal or plastic boxes called pans are used as forms to cast concrete T-beams. (See Figure 17.1.) The pans are set on the falsework in such a position as to form T-beams in one or both directions.

17.4. Parging

Parging is the sealing of masonry surfaces by applying a coat of mortar similar to stucco. Its purpose is to waterproof the area. Parging is applied to catch basins, to manholes, to the outsides of basement walls, and sometimes to the outsides of a cement block wall when they are to receive brick facing.

17.5. Parking Areas

Parking areas are constructed in the same way as roadways. They may be of bituminous or Portland-cement concrete or of a soil–cement mix. The drainage of the surface is very important, so grade stakes must be followed carefully. Excess rainwater on the surface can discourage shoppers from using the area.

Methods of construction are discussed under the various headings.

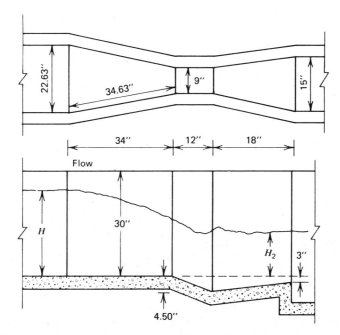

FIGURE 17.2. Parshall flume. The dimensions shown in the illustration are for a flume on an actual project. The throat width is 9 in., but it can be larger or smaller, usually in 3-in. increments. When H_2 of this flume is 6 in., the flow will be 485 gal/min. The flume must be set exactly plumb and horizontal to record correctly. Some are cast out of concrete, and some have a plastic form that is cast into the concrete.

17.6. Parshall Flumes

Figure 17.2 indicates how the flume works. It can be made of concrete or plastic, and temporary ones are made of wood. The flumes are used to measure the flow of sewage or water. Some permanent flumes will have a stilling well on the side with a pipe connected to the flume in the H_2 area so that the liquid level in the wall will be the same as at H_2. A float in the well will provide depth readings for a chart which shows the flow for a period of 24 h or more. The measurements of the flume must be exact, and the walls must be vertical.

Plastic flumes will come as a unit to be set in place and surrounded with concrete.

17.7. Parts Per Million (PPM)

You will see and hear the letters PPM quite often when working on sewage or water treatment plants. The letters stand for "parts per million," as in parts per million of chlorine in water or sewage to purify it. In other words, 1 gal of chlorine to 1 million gal of water.

17.8. Patching

For patching concrete, roads, pipes, roofing, or other items see each under its own heading.

17.9. Paving

Pavement is the hard wearing surface of a road, street, a runway, a taxiway, a parking lot, or any place a hard, durable wearing surface is required. The paving material can be Portland cement or bituminous concrete, water-bound macadam or soil–cement. All surface paving must have an adequate base course or subgrade. (See Sections 3.5 and 20.61.)

Portland-cement concrete will usually have the concrete delivered by Redi-mix trucks, which will deposit the concrete into the hopper of a spreader, or dry batches may be delivered to deposit in the hopper of a mixer. The spreader, which will ride on the forms along the sides of the pour, will distribute the concrete evenly across the width of the pour and screed it to the proper grade. The concrete slab will be 8–10 in. thick, sometimes thicker along the outside edge. The pour will be about 10 ft wide with as many runs as necessary to make the required width of the roadway. Some screeding machines will finish the concrete ready for sealing, while others will level it off, leaving final finish to be done by cement finishers.

The surface will be sealed by spraying on a sealant to retain the moisture longer. The edges will also be sprayed as soon as the forms are removed.

For many years, the standard length of a pour was 50 ft, with an expansion joint at the end of each pour. Now many pours will be over 100 ft long, with reinforcing bars instead of the wire mesh used in the 50 ft runs.

The design of the expansion joint will vary from state to state, but they are all similar, with premolded filler and six to eight expansion bolts for each pour.

Special care should be taken to see that the concrete along the edge and at the expansion joints is well vibrated, to prevent honeycomb.

Some specifications will require that burlap or a sheet of canvas be drawn over the surface just before the concrete sets in order to provide a rough texture for improved traction.

Bituminous concrete mix will be delivered in dump trucks and deposited in the hopper of the spreader. The mix should be hot, between 270° and 290° F. There should be a canvas tarpaulin over the mix in the truck to retain heat and exclude dust. As the mix is deposited in the hopper, a screw conveyor will spread it out to the full width of the hopper. Most spreaders will have a heating element to keep the mix hot until deposited on the subgrade or base course. The loose mix will compact by about one-third when rolled. An inspector will be constantly checking the temperature and thickness of the compacted mix. Sometimes the spreader does not deposit the mix in an even texture, especially when laying a thin course. The uneven spots can be repaired by spreading some mix by hand with a hand shovel. A pail of kerosene should be on hand to dip the shovels in to prevent sticking. Sometimes as the mix is spread an excessive amount of asphalt will show up in spots. These should be removed by square-point shovel, and proper material should be placed before rolling starts.

Shortly after the mix has been laid, the first rolling will be done. This is called the "breakdown" roll. It will partially compact the mix, which is then left to set for a while and cool down, after which the final rolling will continue until the mix is fully compacted to the required density. After each day's run, the equipment must be thoroughly cleaned, and all surfaces that come in contact with the mix should be given a coat of light oil.

The trucks delivering the mix should be spaced to keep the spreader operating full time, but should not allow a bunching up of trucks and a delay in unloading. If the trucks wait too long and the temperature drops in the mix, the inspector may reject the load.

17.10. Payments to the Contractor

Monthly payments to the contractor provide the cash flow to keep the job going. You can play an important part by keeping your work assignment on schedule, with no loose ends waiting to be finished. When the engineer or inspector checks the job to compile an estimate, he will usually hold back funds if there are many partially finished items, such as the backfilling of a trench is incomplete, trash is not cleaned from an area, concrete surfaces are not finished, the backfilling around structures is not complete, or the final coat is not on painted areas. In other words, keep your assignment clean and finished up-to-date as much as possible.

17.11. Payrolls

Prompt payment of wages is expected by most workers, and they get nervous if payment is delayed. You can help by seeing that your time records are complete and up to the minute, with correct names, classifications, wage rates, and hours of regular and overtime. They should be neat and readable when the bookkeeper or accountant comes to collect them.

FIGURE 17.3. Roadway base construction using the penetration method. Four inches of crushed No. 2 and No. 3 stone is used.

17.12. Penetration Macadam

Road building by the penetration method involves placing a mixture of No. 2 and No. 3 crushed stone to a depth of 4–6 in. over the base courses, rolling once to partially compact the stone, and then applying hot asphalt under pressure at a specified rate of gallons per square yard. As soon as the asphalt has been applied, stone chips are spread over the area and the rolling is resumed. The chips are broomed back and forth as the rolling proceeds in order to give a uniform texture to the surface. The rolling will force the chips down between the larger stones to form a tight mass. The rolling will continue until the stones are fully compacted. Some specifications require another application of hot asphalt at fewer gallons per square yard with more stone chips.

Some jobs will have a 1-in. layer of plant-mix bituminous concrete instead of the second application of asphalt and chips.

The hot asphalt will be applied by a tank truck which is equipped to discharge the proper number of gallons per square yard at the required pressure. (See Figure 17.3.)

17.13. Photographs

Progress photos will usually be taken by office personnel; however, it is very useful to be proficient with a camera and take pictures of conditions that occurred on your part of the work. If at a later date a court action requires that you testify, the photos will refresh your memory and satisfy the court that your testimony is correct.

A record should be kept on the back of each photo, listing what is shown, the direction of the camera, where on the site the shot was taken, the time of day, and the date. The negatives should be filed in case additional pictures are required or an enlargement is needed.

If you are working in a built-up area with a great deal of landscaping, photos are recommended as a way of showing what was disturbed and what is required to replace that which was damaged.

17.14. Piers

On civil engineering construction, the word "pier" can mean a platform built out over the water. It may or may not have a structure on it. Such a pier may have a shed for storage in the center, with a platform on one or both sides for loading and unloading ships.

On multispan bridges, the end supports are called abutments and the intermediate supports are piers. (See Section 2.1.)

17.15. Piling

Bearing piling for the support of foundations may be timber (like a telephone pole), reinforced concrete, steel pipe, or structural shapes. It will be specified to carry so many tons, such as a 30 or 50 ton pile. It will support building foundations, bridge piers, retaining walls, and other structures. The piles will be driven by power hammers or sometimes by vibrators. (See Section 9.1.) The tonnage of loading is calculated by several formulas, one of the most popular being the *Engineering News* formula. The formula will give the number of blows per inch for the last few inches for the hammer being used. The hammer will be suspended in the "leads," which will be two heavily reinforced timbers or structural shapes rising vertically some 40 or 50 ft or more, high enough to allow the pile to be placed in the leads under the hammer. (See Figures 17.4–17.7.)

The inspector or engineer will check the pile to see that the diameters of the tip and butt and any bowing of the length of the pile are all within the specified limits. The butt of the pile will be cut so that it is at a right angle to to centerline, allowing the driving force to bear on the whole pile and not just one side. The tip may have a metal shoe attached to prevent crushing of the wood.

A rope or cable will pull the pile up into the leads, and the tip will be set on the spot where the pile is to be driven. The leads will be moved to center the pile vertically within them, and the hammer will be lowered onto

Pile Driving Record

Hammer Data _____   File No: #46 _____

Site ___ LC Shredding Plant ___
Pike, Type† 12" Steel 4 Beam
  Tip Diam. 12 in. Butt Diam 12 in.
  Length 80 ft. ____ in. Weight* _____ lb.
Mandrel (if used)
  Description _____
  Length ____ ft. ____ in. Weight* _____ lb.
Follower (if used)
  Description _____
  Length ____ ft. ____ in. Weight _____ lb.
Penetration:
  Elev. of Ground   _____
  Elev. of Tip (after driving) _____
Time, Start Driving _____ A.M.
                    _____ P.M.

Hammer Make & Model Link Belt _____
  Stroke Rated ____ Meas.†† _____
  Weight of Ram _____ lb.
Driving Cap, Anvil, Helmet, Etc.,
  Weight _____ lb. Description _____
  _____
             (Make sketch on back)
Remarks: φ _____
  Splice @ 40'
  _____
  _____
  _____
  _____
  _____

Finish Driving _____ A.M. _____ P.M.
Driving _____ Mins.

Ft	No. of Blows	Ft	No. of Blows	Ft	No. of Blows	Ft	No. of Blows	Ft	No. of Blows	Ft	No. of Blows	Ft	No. of Blows	Ft	No. of Blows
0		10		20		30		40		50		60		70	
1		11		21		31		41		51		61		71	38
2		12		22		32		42	16	52	21	62	31	72	
3		13		23		33		43		53		63		73	42
4		14		24		34		44		54	24	64	35	74	44 26
5		15		25		35		45	16	55		65		75	
6		16		26		36		46		56	26	66	34	76	48 28
7		17		27		37		47	18	57		67		77	27 35
8		18		28		38	10	48		58	27	68	32	78	43 60 4"
9		19		29		39	10	49	21	59	29	69	38	79	
10		20		30		40		50		60		70		80	

Record number of blows required for each 6 in. of penetration. Note points at which stoppages occur, with times of stopping and starting.

†If wood, state kind, seasoning and treatment. If concrete, state mix and age.
*For wood piles determine actual weight per cubic foot of the wood by weighing a butt section.
††Note any falling off in rated speed and stroke during driving.
φJetting, cause and duration of delays in driving, boulders, bark, condition of cushions, plumbness, banding, damage, driving shoe; etc.

FIGURE 17.4. Record of pile driving. This is a copy of an actual driving record. The notation between 78- and 79-ft level means that 60 blows drove the pile down 4 in.

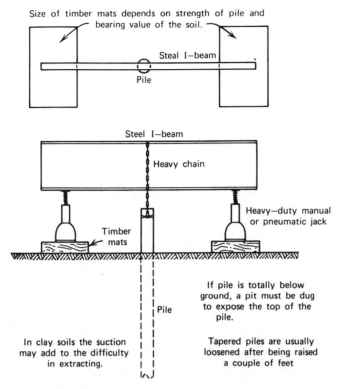

Size of timber mats depends on strength of pile and bearing value of the soil.

Steal I—beam

Pile

Steel I—beam

Heavy chain

Heavy—duty manual or pneumatic jack

Timber mats

Pile

If pile is totally below ground, a pit must be dug to expose the top of the pile.

In clay soils the suction may add to the difficulty in extracting.

Tapered piles are usually loosened after being raised a couple of feet

FIGURE 17.5. Pile puller. A method of pulling pile by use of a steel beam and two heavy-duty jacks is illustrated here. If the pile is totally below the ground, a pit must be dug to expose the top of the pile. Tapered piles usually loosen after being raised 1 or 2 ft. In clay soils, the suction may add to the difficulty in extracting the pile.

the follower, which is a heavy metal and wood device set on top of the pile to prevent the hammer from damaging the butt. As the hammer strikes, the follower will stay on top of the pile as it is driven down. It follows the pile down, hence the name.

At the start of the driving, the hammer will give a few taps to set the pile in the ground, and it will again be checked for verticality before the driving starts. As the driving proceeds, the pile must be observed constantly to see that it remains vertical and to identify any signs that it is being driven off center.

As the hammer reaches full power, the pile will be driven down at a rather constant rate until the tip encounters denser materials or the total friction on the sides of the pile slows it down. At that time, the inspector will start counting the blows per inch of penetration. The driving will be stopped long enough to mark the inches on the side, or the marking may be done before the pile is set in the leads if the depth of driving is known. As the driving continues, the blows per inch for the last few inches will be

FIGURE 17.6. A pile being driven by use of "swinging" leads hanging from the crane cable. Note that the driven piles in the foreground have been cut off to the required elevation, while two in the background near the pile driver have yet to be cut.

FIGURE 17.7. The tops of the foundation piling, which will extend up into the 24-in. reinforced-concrete base slab of a sewage pumping station. Note the heavy H-beam struts holding the steel sheet piling in place. The 8-in. horizontal pipe against the sheeting in the background is the header pipe for the well-point system. The smaller vertical pipes are the well-point riser pipes with heavy reinforced hoses to connect the risers to the header pipe.

recorded, and when the count reaches the required number, the driving stops.

Sometimes, as driving proceeds, the pile will move off center as a result of encountering a boulder or other object. Most specifications will allow a few inches off center. The inspector or engineer will decide. Usually, an off-center pile must be left in place, and a new one must be driven beside it. If a pile in a group is pulled, it will leave a void, which will reduce the friction on the other piles already driven.

Piles carry the load in two ways: by friction on the sides in dense soil and by the tip going through loose soil and bearing on rock or other hard strata. Point-bearing piles can easily be pulled if off center, as the soil friction is not important. One method of pulling piles is shown in Figure 17.5.

Timber piles come in grades A, B, and C, with A being the top grade. The timber piles are more likely to be damaged during driving than the other types. Overdriving can damage the tip or butt or cause the pile to split lengthwise. It will be up to the inspector or engineer to decide if a pile is acceptable; however, it is unwise for a contractor to try to have a questionable pile accepted. Such a pile will probably not carry the full load and could cause settlement cracks in the structure, which the contractor would be required to repair and at considerable expense.

The steel pipe and structural shape (H-column) piles are used when the depth is greater than the length of wood piles available. They are usually point-bearing piles because the friction on their sides is much less than with wood. Such piles are rarely damaged, although overdriving can collapse their tips.

Reinforced concrete piles must be well cured before use, and if so, they are rarely damaged, although again, overdriving can crush their tips.

There is a patented pile in common use that uses a steel shell over a mandrel. The mandrel is tapered and can be reduced in diameter so the steel shell can be slipped over it. The mandrel is then expanded and driven the same as any pile. When the desired loading has been reached, the mandrel is again collapsed and withdrawn, leaving the shell in the ground. On occasion, the mandrel will push a large root to one side, and when the mandrel is withdrawn, the root will push the steel shell in several inches. This can be checked by a flashlight or, if the pile is long, by lowering an electric light. A second shell driven inside the first will sometimes hold the root away until the concrete is poured. If there is water at the bottom of the shell, it should be pumped out before concreting. The concrete should be poured as soon as possible after the shell is driven. Sometimes reinforcing bars will be placed in the upper half of the pile as the concrete is poured.

As the contractor will be paid only for the length of the piling below the cutoff, it is important that a correct record be kept. The inspector will keep a record, but you should also keep your own. Show the length before driving, the length of the cutoff, and the length left in the ground. The length of the last two should equal the first.

Figure 17.4 shows one type of driving record. Such records are usually kept by the inspector, but you should check after each pile is driven to satisfy yourself that the record is correct.

If you and your crew are to be involved in driving timber piles the following procedure is recommended:

1. Have a layout sketch of the piles for correct identification on the driving record.

2. Have the piles inspected as they are unloaded, and set aside any rejects.

3. With chain or cross-cut saw, square off the butt (large) end so that it is perpendicular to the centerline of the pile.

4. Measure the full length of the pile, and mark it in feet and inches on the butt.

5. Shape the tip (small) end to a blunt point, or if a metal shoe is required, shape the tip to fit tightly in the shoe, which will be spiked securely to the pile.

6. Drag the pile to the pile driver with a small tractor.

7. A rope or cable on the pile driver will go over a pulley at the top of the leads and be used to haul the pile up vertically in the leads.

8. As the pile is raised and swung into the leads, remove the stake that marks the location for the pile, and lower the pile so that the tip is in the hole left by the stake. Lower the follower and hammer onto the pile. The weight of the pile, follower, and hammer may cause the pile to settle into the earth, the depth depending on the strength of the soil.

9. Check the pile for verticality both ways, and when vertical, tap it lightly with the hammer, recheck for verticality, and if okay, start the driving.

10. Observe the pile as it is driven to keep it vertical, stopping the driving if necessary to correct it.

11. As the blows per inch increase, stop the driving, mark the inches on the pile, and then resume driving. The inspector will check the blows per inch, and when they reach the required number, he will signal the operator to stop the driving.

12. If the pile is supported by friction on its side, the blows per inch will increase slowly. If the pile is to be point bearing, the blows per inch will increase quickly. Some specifications will require a minimum penetration for the pile, and it will have to be overdriven to reach that depth.

13. If a pile is driven for its full length and the number of blows per inch has not been reached, the engineer will decide if additional piling can be spliced on top or if the pile will have to be pulled and a longer one driven.

14. After a pile has been driven and accepted, the cutoff elevation will be marked on it, and the pile will be cut off at that point. As the cutoff is made, it should be measured, and the length should be recorded. The contractor is usually paid for the length left in the ground, so these measurements are important.

15. When driving in soil with a high clay content, it is not uncommon for the compacted clay to push the pile up several inches. For this reason, the cutting off is usually delayed a day or two, as the pile may have to be redriven. The second driving usually secures the pile.

16. When driving in dense soil, the hammer may suddenly begin to bounce, which usually means the tip has been damaged (broomed). The inspector may require that the pile be pulled and a new one be driven, or he may require that another pile be driven beside it. The pile may encounter a boulder or other hard object, which will push it off vertical and maybe off line. Again, the inspector will decide the action to be taken. Sometimes a pile will split lengthwise and have to be replaced as ordered.

The following method is a quick and easy one for installing steel sheet piling:

1. Be sure the interlocking grooves are clean; if not, clean them with a wire brush or hang them from a crane cable and bang them with a heavy hammer. Have a pail of axle grease on hand, and with a small stick, spread a heavy coating on the bottom few feet of the pile.

2. Mark the line for the piling by sprinkling lime or other white powder on the ground.

3. Set the first sheet pile in position and check it for verticality. Hold it in place by hand or with ropes, set the driving hammer on top, and give the pile a few taps to set it. Again, check for position and verticality. If okay, start driving. Drive the first pile about half the required depth, and then place the second pile and drive it to the same depth.

4. Install five or six sheet piles in this manner, and then go back and drive the first few to the required depth.

5. Continue in this manner, always keeping two or three piles to half depth to serve as guides.

6. If one or more sheet piles hit a boulder or other hard object, leave them up and continue installing piles, driving them to the full depth as soon as the object has been cleared. When the excavation reaches the object, it can be removed, or if it is not in the way, wood sheeting can be installed below.

Figure 17.5 shows an easy way to pull sheet piling, which should be cleaned as soon as it is pulled, while the soil is still moist and will usually fall out if the pile is banged with a hammer. The soil can also be washed out with a garden hose. The groove should be treated with grease or sprayed with oil immediately.

Figure 17.6. Swinging leads are much more maneuverable than the fixed type but they will not control the pile on location as well as the fixed. Swinging leads are particularly useful when driving piling in a basement area where the machine is up on the ground surface. The leads can be spotted anywhere in the basement by extending the crane boom and lowering the leads with the crane cable. The piling shown in Figure 17.7 were driven in this manner.

17.16. Pipe Galleries

In sewage treatment and other processing plants, pipes carry various liquids to many locations throughout the plant. To provide easy access to the pipes, they are routed through galleries where they are exposed for inspection. The gallery may be within a structure or in a concrete tunnel between buildings. Usually, the pipes will have colored arrows showing what is in the pipe and the direction of the flow. Sometimes the pipes are colored.

In underground galleries, especially long ones, there is a problem with leaking when the groundwater table is high. To prevent such leaks, construction joints are placed in the tunnels. These joints will have water stops like those shown in Figure 24.14. Care must be taken to see that the stops are not misaligned during the pouring of the concrete.

There will usually be some moisture on the floor due to sweating of the pipes and condensation on the walls, especially if the groundwater is cold. Most galleries will have floor drains to carry the water to sumps for pumping out.

The concrete must be carefully placed to make it dense and free of honeycomb. A waterproofing of hot asphalt and fabric on the outside may be required. If so, the concrete must be fully dry before the hot asphalt is applied; otherwise, it will not stick. Most jobs will require an application of

FIGURE 17.8. The pipe gallery of a sewage treatment plant under construction. Note in the upper right-hand corner that the bends in the steam line have not been covered with insulation. Later the pipes were color banded with arrows, showing what was in the pipe and the direction of flow.

hot asphalt, a layer of fabric, and then another application of asphalt. The fabric should be well saturated and should overlap several inches at the joints.

Figure 17.8 shows a pipe gallery in a sewage treatment plant designed and supervised by the author. Note that some pipes are suspended from the ceiling, some are set on the floor, and others are in pipe racks along the walls.

17.17. Pipelines

On civil engineering projects, pipelines are used for sewers, the water supply, storm water, and the transporting of oil, gas, air, steam, and all kinds of chemicals and liquids in manufacturing and processing plants. The plans will indicate the kind, size, strength, and alignment of the pipes and whether they are to be buried, supported on the floor, or suspended on hangers. The pipes may be of steel, cast iron, plain or reinforced concrete, clay, asbestos—cement, plastic, or on occasion, wood slats. You must learn the proper method of making the connection between the pipe sections to ensure a tight, nonleaking joint. All pipe should be handled with care, particularly pipe whose ends can be easily damaged to the point where the gaskets will not fit properly.

Pipe sections stored on the site should be placed where equipment cannot run into them and damage the ends. Pipes laid in trenches should have a protective cover of earth over them to prevent damage by stones falling into the trench. Pipes under pressure must have a restraint to keep them from being forced apart at the joints or pushed off line. Those underground may have thrust blocks or tie rods, those above ground may have the same, and long lines may have loops to take the movement. All pipes, whether under or above the ground, should be supported along the body of the section and not by the bells or other connections.

Steel and wrought-iron pipe will usually have flange connections with rubber gaskets. The bolts should be tightened evenly all around. Cast-iron

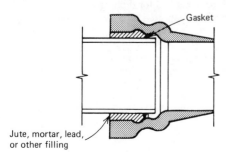

Jute, mortar, lead,
or other filling

FIGURE 17.9. Bell- and spigot- pipe joint.

FIGURE 17.10. A 42-in. reinforced-concrete pipe being laid for drainage of a large airport. The pipe sits on a reinforced-concrete mat, which is supported by 50-ft wood piles because of the very unstable soil in the area. At this point, the pipeline passes under a ridge to get to a creek on the opposite side.

Note that the workers are using a come-along to close the pipe.

When the pipe has been set, forms will be placed and the concrete cradle will be poured.

and concrete pipe will have bell and spigot, while plastic and asbestos–cement pipe will have collar or sleeve connections.

All pipe connections must be clean, and the gasket, if any, should be in place and lubricated when the connection is made. All connections should be up tight.

If the pipe is supported above ground, the supports must be as shown on the plans and placed where indicated so as not to cause too much stress.

See further instructions under the headings for each kind of pipe. See also Figures 17.9 and 17.10.

17.18. Piping

The word "piping" can mean a system of pipes within a structure or on a project. It also refers to a condition in earth dams where a small trickle of water through the dam structure can increase in size and cause the dam to fail. A small trickle of water can carry silt with it, thus enlarging the hole and allowing larger particles to flow out so that the hole gets bigger and bigger until there is a collapse. If discovered soon enough, small trickles can be stopped by digging a short distance into the dam and placing coarse sand in it to filter the silt out. Any piping discovered on the downstream side of a dam should be reported at once to the proper authorities.

If piping is found, filter sand should be placed at once if possible, as a collapse could happen in any time from a few minutes to several days.

It should be noted that some piping in areas containing sand and gravel may bleed for days without any silt being carried out; however, it should be reported.

17.19. Pitot Tubes

A pitot tube is used to measure the flow of water in a stream or from a pipe. Engineers use them to check the flow from hydrants in a new water supply system. A foreman would have little use for the tube, but it is good to know what they are if you see one being used.

17.20. Plan Reading

The ability to read plans is a must for a successful foreman. Most community colleges and many high schools teach plan reading, and if you are not proficient, you should take one of the evening courses. If such schooling is not available to you, there are several correspondence schools that will do as well if you have the ability to study and stick to it until the course is completed. Plan reading is not as difficult as some may think. The wording is usually simple, and the rest is only lines of which you must learn the meaning.

When using plans in stormy weather, a sheet of clear plastic is useful in keeping them dry.

17.21. Plans (See Section 5.16.)

17.22. Plaster

A foreman on a civil engineering project would seldom be involved with plastering. The trade has its own foremen, most of whom came up through the ranks. Labor crews sometimes assist by handling the materials and scaffolding.

For general information, the plaster mix now comes in paper bags, is dumped into a pug mill for mixing (color is sometimes added), and is then conveyed to where the plasterer will use it. The consistency is very important. If it is too wet, the mix will run and not spread; if it is too dry, it will crumble and will not stick to the surface.

There are usually two coats. The first, or "scratch," coat will be grey in color and scratched to provide a bond for the finish coat, which will be a very thin coat of pure white or colored plaster. The finish coat may be troweled smooth to receive paint or wallpaper, or the surface may be roughened to provide a textured finish.

17.23. Plastic Pipe

Plastic pipe, both rigid and flexible, is used for water, sewer, and air lines and for some other liquids and gases. Some rigid pipe is reinforced with fiber for greater strength. The pipe is lightweight and easily handled. Most plastic pipe joints will be a sleeve which fits tightly over the pipe. The pipe ends are treated with an adhesive before the sleeve is placed, and once in place, they adhere to each other, making a water- and airtight joint. Plastic pipe resists corrosion and the buildup of deposits on the inside, much the same as asbestos–cement pipe. Care must be taken to see that the pipe is stored in a safe place, and single lengths should not be left lying on the ground.

When placed underground, plastic pipe should be well bedded under and halfway up the sides. Use only the adhesive compound that comes with the pipe to make the joints. Cover the joint area evenly without excess. It is advisable to wipe the ends of the pipe sections with fine-grade sandpaper before applying the adhesive.

Perforated plastic pipe is used in French drains and sanitary drain fields.

17.24. Plug Valves

Plug valves are used in pipelines carrying heavy liquids such as sewage sludge. The valve inside the housing is a solid unit with an opening in one direction. The valve is turned 90 deg to turn it on or off. When the opening is parallel with the centerline of the pipe, the valve is open, and when turned 90 deg, it is off. These valves are hard to turn, so they must be well lubricated. They will have long pipe handles to give leverage when turning.

17.25. Plumb Bobs

You will find a plumb bob a very useful tool for laying out work, checking vertical lines for forms, excavating, setting falsework, laying sewer and water lines, and many other uses. An inexpensive bob with a holster for your belt can be purchased at most hardware stores.

A heavy plumb bob will be used for deep excavations such as tunnel shafts. The heavy bobs will have wire instead of string. If you are using the bob at considerable height and the wind is a problem, the bob can be suspended in a pail of water, and measurements can be taken from the wire.

17.26. Plumbing

Plumbing is a trade by itself, the work usually being done by subcontract. If you are interested in this work, you should spend the required time as an apprentice worker to learn all the details.

17.27. Pneumatic Caissons (See Section 4.1.)

17.28. Pole Buildings

Originally, pole buildings were, as the name implies, built of poles for framework and covered with rough, undressed planks. One side would be open, like a shed with the opening facing the south and the roof sloping to the north to provide shelter for livestock during storms or winter weather. They were also used to store farm machinery in the off season. The rafters would be poles, with planks or tin sheets for roofing.

Today such buildings are really timber frames, with dressed lumber or corrugated metal sheets for siding.

The original poles were set in holes in the ground, while today most 4 × 4 or 6 × 6 in. verticals are set on concrete footings.

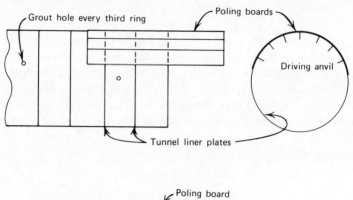

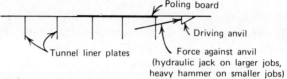

FIGURE 17.11. Poling plates (boards). Note that the poling plates fit outside and above the tunnel liner plates to form a canopy to protect the workers.

17.29. Poling Boards (Plates)

Poling boards are used when tunneling to keep the ceiling from falling in the areas where excavators are working just ahead of the liner plates. (See Figure 17.11.) The boards should extend back over at least two liner rings to give them stability. If the excavated material is earth, it should be left an inch or so low so that the poling boards will shave it off as they are driven forward. This will avoid voids above the liner plates. If the material is shale or rock, it will have to be chipped off as the boards are driven. A heavy (12–14 lb) sledgehammer with a short handle is best for driving the boards forward, unless you have a power hammer. (See Section 21.33.)

17.30. Poly (vinyl chloride) (PVC)

The letters PVC stand for poly (vinyl chloride), a plastic material used to make pipe and tubing and the fittings.

17.31. Posttensioning

Posttensioning, as the name implies, is the tensioning of the reinforcing steel after the concrete has set—just the opposite of prestressing. The steel reinforcing will be placed in tubes, which will be an inch or more larger than the outside diameter of the bars. The bars in the tubes will be placed as indicated on the plans and secured by wire to hold them in place during the pouring of the concrete. After the concrete has attained its full strength, tension will be put on the bars to the amount specified. The steel will be high-strength, the same as for prestressing. The tension will be placed on the bars by an hydraulic jack for large bars and by a nut and torque wrench for small bars. After the tension has been applied, the void between the bar and the tube will be filled by injecting nonshrinking grout. On some projects, the volume of void between the bar and tube will be calculated so it can be ensured that the void is full. Just because some grout comes out the opposite end, that does not ensure that the void is completely filled.

17.32. Pounds Per Square Foot (PSF)

Pounds per square foot is abbreviated PSF.

17.33. Pounds Per Square Inch (PSI)

Pounds per square inch is abbreviated psi.

17.34. Powdermen

The powderman is, or should be, a trained professional, and in most states he will be licensed. He will be responsible for all blasting materials (dynamite, caps, wire, powder). These materials should be stored in a strong shed under lock and key. The powderman will take out only enough material for the day's work and return any not used. He will be in charge of the drilling and will indicate the pattern for the holes. He will personally prepare and place the charges or supervise an assistant. He or his assistant will advise those in the area that the blasting is to take place and sound a warning just before the blast. The warning may be a shout of "fire in the hole" or the blast of a horn or whistle. (See Section 3.19.)

17.35. Power Tools

Power tools such as saws, jackhammers, spades, drills, and vibrators may be electric or compressed-air powered. They make the work easier and faster if the tools are kept in good operating order. Care must be taken to see that the electric cables or air hoses are protected as they lie on the ground. Electric tools should be grounded, with special care taken during wet weather. All tools work better if they are kept properly lubricated and sharp, especially the chain saws. A dull chisel or drill not only works slower, but is more apt to slip off the mark. Not every worker can properly operate all tools, so you must learn who works best with which tool.

17.36. Pozzolith

Pozzolith is added to a concrete mix to make it more plastic without adding more water so that it will more readily flow around the reinforcing bars and into the corners of the forms. The specifications will state if pozzolith is allowed and the quantity limits.

17.37. Precast (Concrete)

Some reinforced concrete structures will be erected by the precast method, which means that the various units that make up the structure, such as columns, beams, girders, and sometimes floor and wall sections, will be cast as individual units, allowed to set to full strength, and then erected similarly to structural steel. The units will be tied together at the connections by welding or bolted steel bars or plates that will protrude out of the ends of each unit. If welding is used, care must be taken to see that the heat from the welding torch does not damage the concrete. If bolts are used, the nuts must be brought up tight without stripping the threads. The use of a torque wrench may be required (see Section 21.22). After the connections have been made, the void will be filled with concrete, which must be well compacted around the steel to prevent rusting. A nonshrinking concrete may be required.

The floor will usually be cast as T-beam sections with one or more beams per unit, depending on the span. The units will be placed side-by-side and may have a finished floor placed over them, depending on the expected use. (See Section 21.5.)

Wall sections will be poured horizontally and, after setting, will be raised to position. They will be a full-story high, with one or more in a panel between columns. The wall sections may be erected before the columns, which will be poured concrete, or if the columns are also precast, there will be steel ties to be welded or bolted.

17.38. Preshooting Rock

When trenching, sometimes shale is encountered that is too hard to excavate by backhoe, but not hard enough to require hard-rock methods. Such areas are preshot by drilling to the elevation of the pipe invert at about 6-ft intervals along the centerline of the proposed line. The blast will loosen the shale to a point where the backhoe can remove it. Such shooting will be done by a licensed or certified powderman, who will decide the amount of the charge and the length of intervals. The power of the blast should be enough to loosen the shale and earth above it, but not enough to cause particles to fly about. Sometimes a layer of hard rock will lie just below the shale so that the blast will send vibrations along the rock to the foundations of nearby structures. The vibrations are seldom enough to cause any damage. Residents in the area should be notified that vibrations may occur and, if so, not to be frightened.

17.39. Pressure Grouting (See Section 8.19.)

17.40. Prestressing

Prestressed concrete will have high-strength (100,000 psi or more) bars, which will be positioned in the forms and fully stressed before the concrete is poured. Special devices at the ends of the forms will keep the bars in tension until the concrete has set. The tension in the bars must not lessen once the concrete pour starts, and the concrete must be well vibrated around the bars to produce a full bond between the concrete and the bars.

Another type of prestressing is used on a round tank that has an unreinforced concrete wall. After the concrete in the wall has fully set, a high-stress wire is wound tightly around the outside of the tank by a special machine that keeps the wire in tension as it winds. After the wire is in place and anchored so that the tension will not lessen, the surface is covered with Gunite to the depth specified. (See Section 8.21.)

17.41. Preventive Maintenance

It is a well known fact that preventive maintenance more than pays for itself by keeping the equipment running and avoiding costly repairs. Usually, a mechanic will be assigned the job of maintenance and will set up a schedule for serving each piece of equipment and supplying it with fuel. Sometimes the schedule breaks down, resulting in a stoppage or slowdown of equipment. If this happens on your work, you should inform your superior at once so the delay will not be charged to you. You should arrange the work of your crew so that when a unit of equipment is stopped for scheduled refueling or servicing, other work can proceed so the workers will not just stand around during the equipment stop.

Preventive maintenance should include checking all points of lubrication and oiling, all cables or hydraulic hoses, wheel tires or caterpillar treads, compressed-air hoses, electric cables, especially the connections, jackhammers, and other power tools. All worn, cracked, or loose parts should be replaced.

17.42. Progress of the Work

All civil engineering project contracts will have a completion date, at which point all work must be competed so the owner may take over. Some contracts will have a problem for liquidated damages if the project runs over the allotted time. These damages can be very expensive for the contractor and should be avoided if possible. You should know how your assignment fits in with the rest of the project so that all sections will move forward on schedule. One small section of the work can delay the whole program if it falls behind its schedule.

Progress depends on production being on schedule, and the skilled foreman makes that possible. The members of his crew must be inspired to work together, equipment must be kept in good operating order, and materials must be on hand when needed. You should be aware of the progress of other crews so that all work goes forward together.

17.43. Project Managers

On large, complicated projects, the owner may hire a project manager at the very inception of the program. He will usually be an engineer with broad experience. He will assist the owner in deciding on the size and makeup of the project, help decide on an engineer or architect, and assist in receiving and analyzing the bids and awarding the contract. During construction, he will assist in expediting the movement of materials and equipment to the job site and try to arbitrate any dispute between the owner and the contractor. He may or may not have a resident engineer under him.

17.44. Proof Rolling

Proof rolling, as the name implies, is the rolling of a subgrade or other surface to prove it is compacted to the required density. When placing bituminous-concrete roadway, the inspector may require the rerolling of a section to satisfy himself that it is okay. When a subgrade has been rolled and is then left exposed for a time, a proof rolling may be required before work can proceed.

17.45. Public Relations

Good public relations is of concern to the contractor, and you are in a position to help. You should be alert to keep noise, dust, and vibration to a minimum. Avoid encroachment on private property and damage to landscaping in the area. If a complaint is made to you, be courteous, even if the complainer is not, and try to resolve the matter. If you cannot, do not ignore the problem, but report it to your superior, explaining what you have tried to do. Many contractors and/or engineers will send out a letter to each local resident, explaining what is to be done and that by the nature of the work some inconvenience will occur, and asking the people to be patient. The things that irritate the people most are noise, dust, obstruction of traffic, hazards to children, and unnecessary damage to property. The contractor or engineer should have notified the fire and police departments, post office, school bus service, and ambulance operators.

17.46. Puddling

Puddling in civil engineering work means using water to consolidate the earth. Backfilling in trenches or around structures can be sped up by spraying with water as the backfill is being placed. The method works best in pervious soils such as run-of-bank gravel and sand. The process is much slower with soils with a high clay content. Such soils will retain the water longer and leave soft areas. As the backfill settles, the water is forced to the top where it must be pumped or drained off. If the surrounding soil is pervious, a good share of the water will leach out during the operation. If the backfill must be consolidated quickly to support loads, the puddling method should not be used.

17.47. Pullshovels

What is now called a backhoe was originally called a pullshovel because the bucket on the boom extends forward and pulls back to load. The bucket operates in just the opposite way from the power shovel, where the bucket moves forward as it loads.

17.48. Pumping Concrete

Originally, concrete pumps were 6 or 8 in. in diameter and used only on projects requiring a large volume of concrete. Now, the smaller-diameter pumps have been developed and are in common use. Pumped concrete is transported from the mixer through the pipe to the point of deposit with no mess along the way, as with wheelbarrows or buggies. There are some problems if the concrete is piped very far. In hot weather, the concrete can set if not kept moving, and it can freeze in winter. The concrete mass must be kept moving once it is in the pipe. At the deposit end of the pipe, there will be a flexible hose, so the concrete can be placed exactly where it is wanted with very little shoveling. On multistory buildings or high bridge decks, the concrete can be pumped vertically several feet and eliminate the need for elevators or other hoisting equipment.

Like all concrete handling equipment, the pump and piping must be cleaned immediately after use. The inside of the pipe must be kept smooth to reduce friction.

17.49. Pumps (Water Removal)

Various types of pumps are used to remove rain and groundwater from a construction site.

Diaphragm pumps (mud suckers) are used to remove water which has accumulated in ditches or trenches. They will remove mud, small stones, and other small objects of up to a couple of inches in diameter. The pump usually has a 4-in. suction and discharge hose. The suction hose is reinforced with steel ribs to prevent collapse from the suction. The rubber or composition diaphragm must be flexible and without cracks or pinholes, which would break the suction. The valves are two rubber balls, one at the suction end and one at the discharge end. These balls must be kept clean so that they will sit tightly. Some diaphragm pumps are hand operated; however, most are motor driven. The suction hose will have a foot valve and a metal screen at the lower end. For the pump to operate efficiently, there should be at least 2 in. of water over the suction end; otherwise, air will get into the line. The diaphragm pumps are self-priming once they are started.

Centrifugal pumps come in almost all sizes, from 1½ to 12 in. or more. The most popular unit is the 2 in. motor-driven pump, which is lightweight and can be easily handled. Some of these are self-priming, but many are not and must be primed each time they are started up. In areas where large amounts of water must be removed, the 4- or 6-in. pumps, and larger, are used. When there is much water to be removed, but not enough to keep a large centrifugal pump running at full capacity, a valve can be placed on the discharge line, and the valve will be closed down until the water discharged equals the amount flowing to the suction end.

For efficient use, all pumps should be serviced regularly and cleaned as soon as they are taken out of service. The suction and discharge hoses should be protected, with particular care to see that trucks or other heavy equipment do not run over them. Heavy planks can be set on both sides to carry the equipment over the hose.

18

18.1. Quality of the Work

Most contractors take pride in doing good, acceptable work. The quality will equal or exceed the norm for the area. To exceed the norm by very much could increase costs to where the contractor would not be competitive when bidding. You should see that the work under your supervision is of a quality that will be accepted, but will not go beyond that unless you are instructed by your superiors to do so.

18.2. Quantity Takeoffs

A construction foreman would seldom, if ever, be involved with a quantity takeoff. However, you might hear a statement such as that the items of work do not agree with the takeoff.

A quantity takeoff is made before the contractor bids on the project. It gives him information on the quantity and quality of each item of work involved in completing the project, including labor, material, and equipment.

18.3. Quarries

A quarry is an excavation where both cut and crushed stone are obtained for construction projects.

The cut stone will be in square or rectangular blocks, and the crushed stone will be broken up in a crusher to sizes from ½ in. to 4 in. or more, depending on the use intended. The stone will be obtained by drilling and blasting in a pattern designed to loosen the stone to the approximate size needed. The large pieces will be cut by a carborundum saw to the shapes required, and the smaller ones will be crushed down to the sizes needed. The cut pieces will be shipped to a mill for final shaping and polishing. The crushed stone will be used as the base course for a road, for concrete, or for another such use.

18.4. Quicksand

Quicksand is a condition of the soil, not a type of sand. If there is a high level of groundwater in the area, you should be alert for a quicksand condition as excavation proceeds. If the soil is pervious, as are sand and gravel, there will be boils without a quicksand condition. (See Section 3.22.) If there is a high clay content in the soil, the groundwater will pass through much more slowly and saturate the fine particles of silt and clay, causing a jelly-like mass and a quicksand condition. As the excavation continues, you will note that the soil has a rubbery condition, and it will move slightly if you tap it with your foot. Continue to tap, and water will usually show up on the surface. If you encounter such a condition, especially if the plans and/ or specifications do not mention quicksand, you should notify your superior at once, as a claim for extra work may be justified. Under such conditions, you should keep a record of all costs of labor, materials, and equipment time that could be attributed to the delay.

A quicksand condition can be made stable by removing the groundwater, usually by well points, and consolidating the soil by vibration. This is a costly operation, and unless the contract documents indicated that quicksand might be encountered, there should be justification for a claim.

18.5. Quick-Set Cements

Quick-set cement is used in the mortar for stopping leaks in concrete, under column-bearing plates, under the base of permanent equipment, and any-place where the mortar must not shrink. (See Section 15.5.)

There are several brands on the market, and although they are similar, you should follow the instructions on the container.

Some quick-set cements generate considerable heat, so heavy rubber gloves should be worn. The mortar should be mixed in small batches and used at once, and the gloves should be washed after each batch.

19

19.1. Railroad Construction

New railroad construction within the United States is practically zero. Roadbeds are upgraded with new ballast and rails and regular maintenance, and that is about all.

In some foreign countries, it is a different story, with new construction and the upgrading and extending of lines. In case you are interested in overseas work, common procedure is as follows. On such work, many construction foremen start with the survey crews, clearing brush and trees and assisting with the survey equipment. After a route has been established, full clearing of the right-of-way is next. The trees and brush including the roots, are removed to prevent further growth, and burned or otherwise destroyed. Then excavation begins to establish the desired grade. Hills are cut down, and valleys are filled. Culverts or bridges are built in the valleys. The embankment fill is compacted as it is placed, and the slope is carried out to the design requirements. The cut and fill will be brought to the design grade. Then one or more feet of ballast is placed to the height of the bottom of the ties. The rails will be placed and secured to the ties, being set to the proper line and grade. Additional ballast is then placed between and around the ties to hold them in place. The ballast will be placed and compacted to the top of the ties.

Switches and sidetracks will be placed as required. Additional clearing will be done at all highway crossings to give the required sight distance.

The culverts can be concrete, stone, or metal and, in some remote areas, might be timber. The bridges will be timber, steel, or concrete, or a combination of these.

FIGURE 19.1. Old-style railroad track. Old-style tracks had 33-ft-long sections of rail connected by fish plates. A jumper cable was required at each joint to carry the signal messages. Note the tie plates under the rails at each tie, with spikes through the plate into the tie.

FIGURE 19.2. New-style railroad tracks. New-style rails may be up to 200 ft long with welded joints. The welds are ground down to make a smooth joint, thus eliminating most of the wheel noises.

The centerline of the tracks will be set from tacks on the control stakes set about 50 ft on centers. The elevation of the base of the rails will be set from the tops of the control stakes. On a curve, the base of the lower rail is set to grade, and the higher one is set to whatever superelevation is required. Construction methods for all other work are discussed elsewhere in the book. (See Figures 19.1 and 19.2.)

19.2. Railroads (Marine)

Marine railroad is the name given to rails on which boats and ships are launched. The rails will run from an area of dry land where the boat or ship is built or stored out into the water as far as necessary to launch the size boat or ship involved.

There can be two or more rails set on ties of whatever length is required. On land, the tops of the ties may be at or above the surface of the ground. Underwater, they are usually set on the bottom and covered with stone ballast up to or above the base of the rails. When large ships are to be launched, the underwater rails may be set in concrete, which is placed by dewatering the area inside a cofferdam. When the railroad is for small boats, the standard ties and lightweight rails are assembled on land and dragged into the water and weighted down with stone ballast.

19.3. Records (Reports)

Record keeping by the foreman varies from one contractor to another, and instructions will be given on what is required. Some of the records required are as follows:

1. *Timekeeping.* Items required are full name, social security number, labor classification, and hours worked at various tasks, with a summary of total hours for the day. Overtime work hours are kept separately from regular hours.

2. *Equipment Used.* List all hand tools and large equipment used, with the exact hours for each piece of equipment and what it was doing. List any hand tools broken or damaged, and any equipment problems, with time out for repairs or servicing.

3. *Materials Used.* Describe and quantify all materials used, with notes on any damaged material.

4. *Official Visitors.* List any official visitors and any conversations or instructions. Note the time of the visit.

5. *Weather.* Note weather conditions, good or bad, especially weather that disrupted work. If disrupted, note the time you were delayed.

Most contractors will have their own books for keeping records, and they will be supplied to the foreman as required. (See Figures 19.3 and 19.4.)

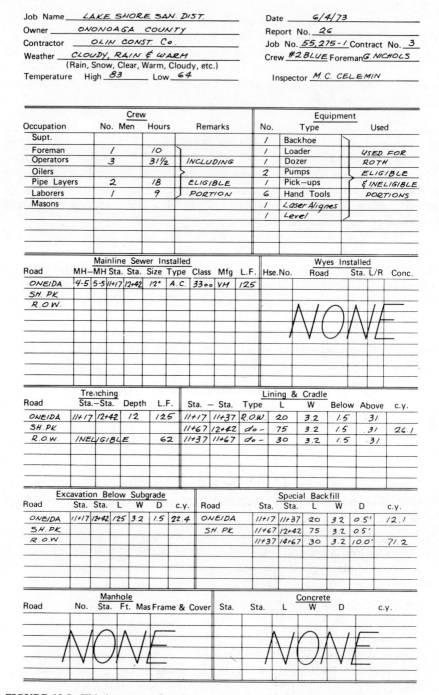

Job Name ___LAKE SHORE SAN. DIST.___ Date ___6/4/73___
Owner ___ONONOAGA COUNTY___ Report No. _26_
Contractor ___OLIN CONST. Co.___ Job No. _55,275-1_ Contract No. _3_
Weather ___CLOUDY, RAIN & WARM___ Crew _#2 BLUE_ Foreman _G. NICHOLS_
(Rain, Snow, Clear, Warm, Cloudy, etc.)
Temperature High _83_ Low _64_ Inspector _M.C. CELEMIN_

Crew

Occupation	No. Men	Hours	Remarks
Supt.			
Foreman	1	10	
Operators	3	31½	INCLUDING
Oilers			
Pipe Layers	2	18	ELIGIBLE
Laborers	1	9	PORTION
Masons			

Equipment

No.	Type	Used
1	Backhoe	
1	Loader	USED FOR
1	Dozer	ROTH
2	Pumps	ELIGIBLE
1	Pick-ups	& INELIGIBLE
6	Hand Tools	PORTIONS
1	Laser Alignes	
1	Level	

Mainline Sewer Installed

Road	MH–MH	Sta.	Sta.	Size	Type	Class	Mfg	L.F.
ONEIDA	4-5	5-5	11+17	12+42	12"	A.C.	3300	VM	125
SH. PK								
R.O.W.								

Wyes Installed

NONE

Trenching

Road	Sta.–Sta.	Depth	L.F.
ONEIDA	11+17	12+42	12	125
SH.PK			
R.O.W.	INELIGIBLE		62

Lining & Cradle

Sta. – Sta.	Type	L	W	Below	Above	c.y.
11+17	11+37	R.O.W.	20	3.2	1.5	.31	
11+67	12+42	do –	75	3.2	1.5	.31	26.1
11+37	11+67	do –	30	3.2	1.5	.31	

Excavation Below Subgrade

Road	Sta.	Sta.	L	W	D	c.y.
ONEIDA	11+17	12+42	125	3.2	1.5	22.4
SH.PK						
R.O.W.						

Special Backfill

Road	Sta.	Sta.	L	W	D	c.y.
ONEIDA	11+17	11+37	20	3.2	0.5'	12.1
SH.PK	11+67	12+42	75	3.2	0.5'	
	11+37	14+67	30	3.2	10.0'	71.2

Manhole

Road	No.	Sta.	Ft.	Mas	Frame & Cover

NONE

Concrete

Sta.	Sta.	L	W	D	c.y.

NONE

FIGURE 19.3. This is a copy of an inspector's report. A foreman's report gives the same information to the superintendent.

Consulting Engineers and Land Surveyors

Type of Project: PUMPING STATION

Job Name Polar Beech P.S. Job No. 55.37

Owner ________________ Report No. 719

Contractor ________________ Date 7-7-77

Weather: Fair Aw/Pa Rain____ Warm____
Snow____ Cold____ Very Cold____
Temp: High° 80 Low° 60

Inspector RAK

Occupation	No.	Total Men	Total Hours	Remarks
Carpenters	3		8	
Laborers	4		8	
Finisher	1		8	

Equipment: Crane - Concrete Bucket

Work Accomplished:
- Placing re-bars in int/vent channel area
- Set Pair Shall flume in place
- Poured floor in int/vent channel area to ele. 357.50 ± - Finished int/vent channel itself. 33 cy ±

FIGURE 19.4. Inspection records. Shown here are copies of two types of daily report form that can be used by construction inspectors. A foreman's report gives the same information to the superintendent.

19.4. Redi-Mix (Concrete) (See Section 4.35.)

19.5. Reinforcing Steel

The position of the reinforcing bars in a concrete beam, girder, column, or slab is very important. They should be secured so that they cannot move during the pouring of the concrete. A change in position could change the strength considerably. The plans will show the positions the designer intended. All reinforcing should be free from rust scales, dirt, oil, or grease. If the

FIGURE 19.5. Reinforcing steel bars used for the base or floor slab of a circular tank. Dowels are set for the wall's reinforcing steel. Note that concrete pouring starts at the top of the slope and works toward the center or lower area. (bottom) The worker with the cable over his shoulder is the vibrator operator. Note the concrete bucket suspended from the crane cable.

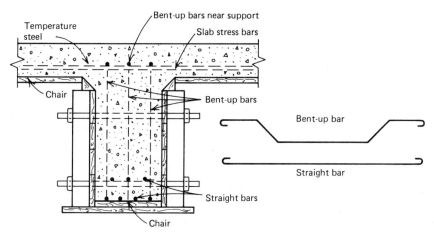

FIGURE 19.6. Reinforcing bar.

FIGURE 19.7. Reinforcing bars. Close up view of the slab shown in Figure 19.5. Note that the circular bars near the center have not yet been tied in place. The forms around the perimeter will provide a horizontal base for the wall. The photo gives an idea of the difficulty of pouring and vibrating the concrete in place through the steel bars. This slab will be 18 in. thick.

steel has been stockpiled on the site, it may be necessary to clean it with a wire brush before installation.

To keep the bars in position during the pouring of the concrete, most contractors will use metal chairs designed for that purpose (see Section 4.18). The bent-up bars may be wired to the stirrup bars or the slab steel (see Section 20.53). Figures 19.5–19.7 indicate how the bars are positioned.

The bars in concrete slabs may be placed in both directions or in only one direction, with smaller temperature bars in the opposite direction, the temperature bars being on top. In either case, the bars should be wired together at each intersection.

Most reinforcing bars will have a maximum length of 40 ft, with splicing required for longer lengths. The specifications will indicate the length of the splice required for the various diameter bars, usually 30 diameters.

A reinforcing bar is numbered by the number of eigths of inches in its diameter. Thus, a No. 2 bar will be ¼ in. in diameter, a No. 3 bar will be ⅜ in., a No. 4 bar will be ½ in., and so on.

Concrete slabs for floors and highways will usually have a wire mesh for reinforcing. The gauge of the wire will depend on the loads on the slab. The heavy-gauge mesh will come in sheets, and the lighter gauges will come in rolls. Mesh reinforcing will usually be placed at the center of the slab; however, it is best to check the plans to be sure.

19.6. Rejected Work or Equipment

In spite of the best efforts, sometimes a portion of the work or a unit of the permanent equipment will be rejected. Such action usually reflects on the foreman, whether it is his fault or not. You must see that all members of your crew are skilled in the work at hand, and if not, that they have constant supervision until they learn. Units of the permanent equipment that are handled and installed by your crew must, when temporarily stored on the site, be protected from damage. When they are handled for installation, each unit must be properly secured so as not to slip, fall, or strike another object.

Concrete poured in freezing weather must be protected by heating equipment that will not fail. Freshly painted areas must be protected from dust and other debris. Forms for concrete must be secure so as not to move until the concrete has set. Wavy lines are unsightly and costly to remove.

19.7. Repairs

You would not normally think about repairs on a project still under construction, but it does happen. Just some of the reasons why repairs are required on jobs still underway are: improper construction methods; damage to permanent equipment due to dropping or damage while in storage on the site or due to improper installation; misalignment of concrete forms or settling while pouring; settlement of the subgrade in an area that was not properly checked; or a leak in a pipe that was not properly checked. The repairs are costly to the contractor and can be avoided by good supervision.

If some of the permanent equipment being installed on the project is damaged, it is usually required that the repairs be made by a factory-trained mechanic. The unit may be repaired on site or may require replacement.

If concrete forms become misaligned, it is sometimes possible to correct them as the concreting continues. If the concrete has set, it is more difficult. A warp in a concrete wall or beam may be corrected by chipping off and refinishing the surface, or it may be necessary to roughen the surface for bond and to build out the area. If a beam or girder has settled during pouring and has a reverse camber, it will probably have to be replaced. All honeycomb has to be cut out, and the area must be repaired.

If a pavement has settled due to a subgrade failure, the whole area will have to be dug up down through the soft material to solid earth, the area will have to be backfilled with properly compacted materials, and the paving will have to be replaced.

Leaks in pipes may sometimes be corrected by installing a collar over a crack, and leaks at the bell can be corrected by a bell clamp. Leaks in a poured lead joint may be fixed by use of a caulking iron and hammer.

Leaks in built-up roofing can be repaired by removing the top layer, applying hot tar and fabric, and sealing the joint between the old and new with hot tar.

Damaged brick or tile can be repaired by chipping out the damaged unit and replacing it with a new unit that matches.

Painted surfaces that have become dirty or have been scratched should be repainted for the full height to a break in the wall—one whole side of a room might be repainted, for example.

Ceramic tile can be repaired by chipping out the damaged unit and about ½ in. of the backup mortar. Apply fresh mortar and install a new matching tile.

Terrazzo floors that become damaged will require the removal of a whole section. If the floor has been installed in 3- or 4-ft squares with metal dividers, only the damaged square need be removed. However, it is difficult to match a terrazzo surface that has been in place for awhile.

19.8. Resident Engineers

Civil engineering projects will have a resident engineer in charge for the owner. He will be a trained professional and probably licensed. He will have one or more inspectors working under him, depending on the size and scope of the project. His job is to see that all work on the contract is completed as called for by the plans and specifications and to see that the contractor observes all laws and regulations pertaining to the work.

The resident engineer will compile or review the monthly estimate of work done and okay payment to the contractor.

The specifications will detail his responsibilities and state the limitations, if any, on his authority.

If the engineer criticizes the work of your crew, do not argue with him, but confer with your superior and get the problem straightened out as quickly as possible to avoid delay.

19.9. Retaining Walls

Retaining walls, as the name implies, retain the earth at a higher elevation, as shown in Figure 19.8. Usually reinforced concrete is used, but some are large quarried stone set with or without mortar. The stones must be large enough to resist the pressure from the earth. Drainage and weepholes are necessary, as shown.

19.10. Revetments

A revetment is a low, usually temporary, retaining wall. It can be made of sandbags, earth, timber, or masonry.

19.11. Revolutions Per Minute (rpm)

The letters rpm stand for revolutions per minute and indicate the speed at which a machine is operating. For example, a pump must operate at a given rpm to discharge a certain number of gallons per minute (gpm). An engine must operate at a given rpm to produce a required horsepower. Many machines will have a gauge on the unit which will indicate the rpm. Also, there is a device that can be pressed against the end of a rotating shaft to indicate the rpm.

19.12. Rights-of-Way

A right-of-way is quite often required for a water or sewer line. It is supposed to be acquired before the contact is let and should not be a problem for the contractor. However, sometimes there is a slip-up and the right-of-way has not been obtained at the time the contractor is ready to proceed or even underway. Some property owners will not allow the work to proceed

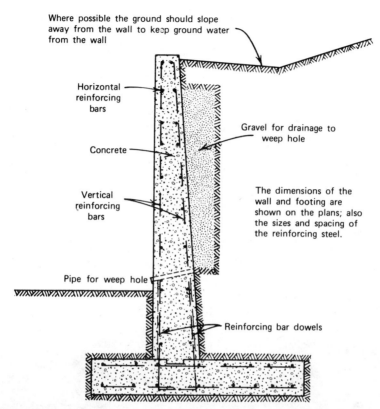

Where possible the ground should slope away from the wall to keep ground water from the wall

Horizontal reinforcing bars

Concrete

Vertical reinforcing bars

Pipe for weep hole

Gravel for drainage to weep hole

The dimensions of the wall and footing are shown on the plans; also the sizes and spacing of the reinforcing steel.

Reinforcing bar dowels

FIGURE 19.8. Retaining wall. Note the gravel for draining behind the wall with a weep hole to carry off the water. The dimensions of the wall and footing, as well as the size and spacing of the reinforcing bars, are shown on the plans. When possible, the ground should slope away from the wall as shown to keep the surface and groundwater away.

over their land until the right-of-way has been acquired. Under such circumstances, the contractor's work may be delayed, and he will be entitled to payment for the costs of the delay. If your crew is delayed by such problems, you should keep accurate records of all time lost by workers and equipment.

Working within a right-of-way can restrict the area of operations and can result in damage claims if adjacent property is disturbed. You should know the limitations, and inform your crew, and see that they stay within them.

19.13. Riprap

Riprap is used to prevent erosion along river banks and embankments leading to bridges, in drainage ditches, and at other places where moving water can erode the earth. (See Figure 19.9.) The material can be broken

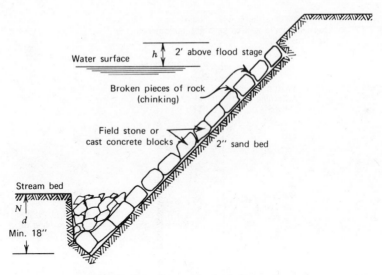

FIGURE 19.9. One method of placing riprap.

FIGURE 19.10. "Dumped" riprap along a lakeshore being used to provide protection from wave action during windstorms. The stone was dumped from trucks and spread by dozer. Note the concrete head wall for the bypass pipe from the sewage pumping station on the left.

FIGURE 19.11. "Dumped" riprap on the upstream slope of an earthen dam for a flood-controlled basin. The downstream slope is grass covered. The concrete spillway prevents overtopping of the dam at high flood stage.

The flow in the small creek passes under the dam through a 30-in. pipe with a valve that is closed during flooding. The small structure at the top of the dam contains the controls for operating the valve.

FIGURE 19.12. Dumped riprap along the bottom of the slope in a "cut" area. This stops erosion from groundwater bleeding from the slope.

stone, broken pieces of old concrete slabs, or sacks filled with concrete. The sacks can be jammed tightly together to make a solid surface, but all the others will require chinking. That is, small pieces of stone or concrete must be driven into the joints between the material to make a tight surface.

The stone or broken concrete can be dumped on the slope in small pieces to a depth of several inches, and with a small amount of hand leveling, that will work okay. The larger pieces will be arranged by hand to present as tight a surface as possible and will then be chinked. On some projects, the large stones are arranged in a pattern by hand and then chinked, but this is expensive and rarely done.

Note that in Figure 19.9 the riprap is carried down below the riverbed to prevent scouring. In deep water, such as in a reservoir, the riprap may extend from a few feet above high water to a few feet below low water to prevent erosion by wave action (see Section 8.1). (See Figures 19.10–19.12.)

19.14. River Crossings

Constructing pipelines across rivers or larger creeks can present problems if there is boat traffic on the water or if the waters hold game fish.

Rope or cable to anchors for equipment must have warning signs of some kind to keep boats from tangling their propellers. Floating equipment must be anchored so as not to swing in the wind and block the traffic lane. Oil or gasoline spills must not be large enough to kill the fish. The backfill of a trench should not be higher than the river bottom, or if higher, it should be low enough below the surface to prevent scouring from the propeller wash.

Crossing for a sewer line is done by an inverted siphon, unless the sewer grade is low enough to be below the river bottom. Water lines should be at least 2 ft below the bottom, and more if large ships cause heavy propeller wash.

Crossings over nonnavigable waters can be above the water surface on piers with or without bridgework. In areas of freezing weather, the water or sewer pipe must be insulated. This is usually done by wrapping a blanket of insulation around the pipe and covering it with metal or fiber as protection from damage. Force mains are treated the same as water mains.

19.15. Roads (Streets)

Roads are classified as arterial (such as Interstate and U.S. highways), primary (such as state highways), secondary (such as county highways), access roads (to the first three types and to parallel service roads), and farm-to-market roads (the rural roads leading to villages and to the others noted above).

The arterial roads which must carry heavy trucks at high speed will have a surface course of reinforced concrete, bituminous concrete, or a combination of these. Unless there is an unusually firm subgrade, there will be several layers of base course, which could be 3 ft or more in thickness. Each layer of crushed stone, usually No. 2 and No. 3 mixed, will be thoroughly compacted before the next layer is placed. The bottom layers could be run-of-bank gravel, and the top layer could be penetrated with hot asphalt. The plans will indicate the type and material of each layer. Because a pavement failure under such heavy, high-speed traffic could cause serious accidents, the subgrade and each layer of the highway must be the proper material and thoroughly compacted. Unless there is natural drainage, provision must be made to carry off the water that accumulates along the subgrade and base courses. Where there is a considerable amount of water, French drains or another such method will be used to carry the water away from the highway.

U.S. highways will be constructed very similarly to the arterials.

The primary or state highways are also constructed similarly to the arterials, only usually not to quite as high a standard. Many states will not permit on their highways trucks of the weight and length of those using the arterials.

Farm-to-market roads are the rural roads that have been upgraded by adding run-of-bank gravel, been shaped up to form a crown, and had the ditches opened to provide better drainage. They may have a surface treatment of oil and stone. (See Sections 3.5, 3.18, 17.9, and 20.24.)

19.16. Rock Bolts

Rock bolts are used to hold loose rock layers in place in tunnels or rock cliffs along highways. When there is a danger that some rock may loosen and fall, holes are drilled through the loose layers into the solid rock behind. Steel bolts with expanding devices on the ends are inserted into the holes and expanded to secure the bolts in place, and then the holes will be filled with grout to protect the bolts from rusting. Figure 19.13 shows how the bolts are used. The roof of a tunnel may be arched or flat. Rock bolts are also used with tiebacks. (See Figure 21.10.)

19.17. Rolling (Compacting)

Compacting by rolling is done for subgrades, embankments, backfill, and bituminous-concrete paving. There are several types of rollers: the three-wheeled old standard with one roller in front and one on each side in the rear, the three-wheeled tandem, the two-wheeled tandem, the sheepsfoot roller, and the rubber-tired roller. There are modifications of these, such as the steel-wheel vibrating roller with two rubber-tired wheels in back. The first three types can be used for any rolling; the sheepsfoot and rubber-tired rollers are used on soil with a high clay content.

On bituminous paving, the rolling will usually start with the old standard and finish with a two- or three-wheeled tandem to iron out the ridges.

Extra weight can be added to the steel-wheeled rollers by putting water inside the roller. On sheepsfoot or rubber-tired rollers, the weight is added by placing sand bags or other weights on top. On the steel-wheeled rollers, there will be a steel bar at the top to scrape off any earth or other substance that may stick on the roller. While rolling bituminous concrete, there will be a spray of water over the roller to prevent sticking. On all rolling, each run should overlap the previous run by 50%.

When compacting soil with a high clay content, the sheepsfoot roller will be used. At first, the "feet" will penetrate the soil for its full depth, and as the soil is compacted, the feet will penetrate less, until at last they will be on top with very little penetration. Then the rubber-tired roller is used to complete the compacting. Many specifications state the tire pressure that must be used. (See Section 4.32.)

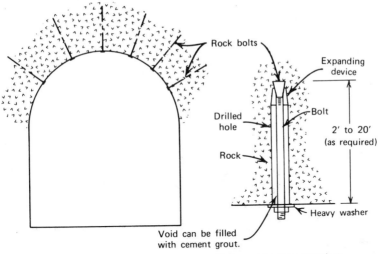

FIGURE 19.13. Rock bolts used in a tunnel. The bolts must be long enough to expand in solid rock. The size of the bolt and the expanding devide depend on the load to be sustained. The specifications indicate the sizes and lengths. The tunnel foreman relies on his judgments as to spacing.

FIGURE 19.14. Rubber-tired roller with water tank for increasing the weight. Note the coupling device on the rear of the vehicle. This is used to tow a steel-wheeled or sheep's-foot roller. The tires on these rollers have smooth treads.

19.18. Roofing

Several types of roofing are used, depending on the type of structure and its use.

Most civil engineering projects will have flat roofs with built-up roofing. First, a coating of hot tar is spread over the roof boards and a layer of fabric is rolled over it. The fabric will be pressed down into the tar to become saturated. Another coating of hot tar is spread, with another layer of fabric. The specifications will state the number of layers of fabric required. The hot tar will be spread over the last layer of fabric. Some specifications will require a layer of small pebbles over the top coat of tar as a wearing surface.

If the roof ends against a wall, there will be a recess in the wall, with the roofing terminating up in it (see Figure 7.8). When the roof extends beyond the wall as an overhang, the edge will have a metal gravel stop to hold the gravel and keep the ends of the roofing fabric from raveling.

Built-up roofs are not exactly flat. They will have enough slope to carry the rainwater to the edge of the roofing or to roof drains on the surface.

Occasionally, there will be a sloping or gabled roof. These will have tile, shingles, or metal sheets for roofing. The roof boards will be covered with a layer of roofing paper nailed in place, and the tiles will be set on a gob of mortar. Shingles will be nailed in place with an overlap of about a quarter of their width. Metal sheets will be nailed to roof boards or wood strips set several inches apart.

Any chimneys, vent pipes, or roof drains will have flashing to prevent leaks. (See Figures 7.6–7.8.)

A new material added to roofing projects in recent years is sheet rubber. The sheets can be applied over roofboards or insulation. The joints are sealed to provide a watertight covering. Usually, a layer of small, washed gravel is placed over the sheets as a protective covering. The rubber is tough, not easily damaged, and easily repaired.

19.19. Room-Finish Schedules

A building with finished wall surfaces inside will have a room-finish schedule on the plans. The rooms will each have a number for reference to the schedule. Figure 19.14 will show you how the schedule works.

19.20. Room Separators

Convention centers and large hotels will have a long room for large banquets or conventions. The room will have accordion or folding partitions by which the large room can be made into several smaller ones. The partitions will

ROOM FINISH SCHEDULE

ROOM NO.	NAME	FLOOR	BASE	WALLS				CEILING		REMARKS
				NORTH	EAST	SOUTH	WEST	MAT'L	HEIGHT	
100	VESTIBULE	FL. MAT	VINYL	MAS. PTD	—	MAS. PTD	V. W. C.	ACUS. TILE	9'0"	MAS. PTD – MASONRY PAINTED
101	LOBBY	CARPET	VINYL	MAS. PTD V.W.C.	MAS. PTD V.W.C.	MAS. PTD	V. W. C.	ACUS. TILE	9'0"	V.W.C. – VINYL WALL COVERING
102	OFFICE	CARPET	VINYL	V.W.C./GB	V.W.C./GB	V.W.C./GB	V.W.C./GB	ACUS. TILE	9'0"	ACUS. TILE – ACOUSTICAL TILE
103	CONFERENCE	CARPET	VINYL	V.W.C./GB	V.W.C./GB	V.W.C./GB	V.W.C./GB	ACUS. TILE	9'0"	V.W.C./GB – VINYL WALL COVERING ON GYPSUM BOARD
104	REST ROOM	C.T.	C.T.	C.T.	C.T.	C.T.	C.T.	ACUS. TILE	9'0"	C.T. – CERAMIC TILE
105	LABORATORY	V.A.T.	VINYL	V.W.C./GB	V.W.C./GB	V.W.C./GB	MAS. PTD	ACUS. TILE	9'0"	
106	ACCESS ROOM	CONC. PTD	—	MAS. PTD	MAS. PTD	MAS. PTD	MAS. PTD	CONC. PTD		CONC. PTD – CONCRETE PAINTED
107	CHLORINATION	CONC. PTD	—	MAS. PTD	MAS. PTD	MAS. PTD	MAS. PTD	CONC. PTD		
108	ELECTRICAL ROOM	V.A.T.	VINYL	MAS. PTD	MAS. PTD	MAS. PTD	MAS. PTD	CONC. PTD		V.A.T. – VINYL ASBESTOS TILE
109	BLOWER ROOM	CARPET	VINYL	MAS. PTD	MAS. PTD	MAS. PTD	MAS. PTD	TECTUM TILE	9'0"	ACUS. PANELS ON WALLS
110	GENERATOR ROOM	CONC.	—	MAS. PTD	MAS. PTD	MAS. PTD	MAS. PTD	CONC. PTD		
111	GARAGE	CONC.	—	MAS. PTD	MAS. PTD	MAS. PTD	MAS. PTD	CONC. PTD		CONC. – CONCRETE
112	STORAGE	CONC.	—	MAS. PTD	MAS. PTD	MAS. PTD	MAS. PTD	CONC. PTD		
113	WORKSHOP	V.A.T.	VINYL	MAS. PTD	MAS. PTD	MAS. PTD	MAS. PTD	CONC. PTD		
114	STORAGE	V.A.T.	VINYL	MAS. PTD	MAS. PTD	MAS. PTD	MAS. PTD	CONC. PTD		
115	JANITOR	V.A.T.	VINYL	MAS. PTD	MAS. PTD	MAS. PTD	MAS. PTD	CONC. PTD		
116	CORRIDOR	V.A.T.	VINYL	MAS. PTD	—	MAS. PTD	—	ACUS. TILE	9'0"	
117	CLOSET	CARPET	VINYL	V.W.C./GB	V.W.C./GB	V.W.C./GB	VWC/GB	ACUS. TILE	9'0"	

ABBREVIATIONS: V.A.T. – Vinyl Asbestos Tile; V.W.C. – Vinyl Wall Covering; C.T. – Ceramic Tile; CONC. – Concrete; CONC. PTD – Concrete Painted; MAS. PTD – Masonry Painted; ACUS. TILE – Acoustical Tile; V.W.C./GB – Vinyl Wall Covering on Gypsum Board.

FIGURE 19.15. Room-finish schedule.

297

be suspended from the ceiling on rollers running in a track in the ceiling. Some units will also have rollers running in a track on the floor for more stability. The units will either fold back into a recess in the wall or just fold against the wall.

While the units will separate the area into smaller rooms, they have little or no insulation, so sound passes easily from room to room.

19.21. Roughing-in (See Section 17.26.)

19.22. Rubbing (Concrete)

Rubbing is a treatment of the surface of concrete to fill the pinholes and other imperfections and present a uniform texture. The rubbing is done by use of a carborundum stone and a small amount of water. The water is sprinkled over the area, which is then rubbed with the stone using a circular movement. The rubbing may be done by hand or machine. The rubbing will work up a lather of mortar, which is worked into the pinholes and other areas by use of a damp brush or a piece of damp burlap. The lather should be worked as little as possible to get the surface smooth. Too much working will cause the surface to become dusty when dried out. Too thick a layer of lather left on the surface will result in thin hair cracks, like a spider's web, unless dried out very slowly.

If rubbing is required, the specifications will indicate how it is to be done.

19.23. Runways (Airport)

Airport runways, the surfaces on which the planes take off and land, are constructed similarly to highways, except that the runways are much thicker in order to take the impact of landing. The thickness and width of the pavement will depend on the class of airport. Large commercial airports will have runways several feet thick and several hundred feet wide. The surfaces may be concrete or bituminous; however, the bituminous can be cut as the wheel touches down. The plans will indicate the design to be used. As the impact of the wheels is several hundred pounds per square inch, the drainage of the subgrade is important, and usually special provisions are made to keep it dry. Many runways will have a thick base course of several layers of bituminous concrete and a wearing surface of reinforced concrete. The methods of installing the various courses will be indicated on the plans and in the specifications. As with highways, the runway is no better than the subgrade on which it is built.

19.24. Rust

Painting over rust on metal can be costly for the contractor if the contract has a one-year guarantee. Loose scales or rust covered with paint will continue to expand and push the paint up. To correct the problem, the paint has to be removed along with the rust, down to bright metal. Then the area has to be primed and the required number of coats of paint must be applied. Also, it is sometimes hard to match the paint exactly.

Rust can be removed by hand or power-operated steel wire brush or by sand or carborundum paper.

All loose scales and rust should be removed, and the area should be wiped clean of all dust.

Welding will destroy the paint for several inches on each side of the weld, and the area will start to rust quickly. Such areas should be cleaned and repainted as soon as possible.

20

20.1. Safety

The promotion of safety habits on a construction project is an ever-present chore. It affects the morale of the workers and their production. It helps keep insurance rates lower and reduces the cost of replacing damaged items. Many highly skilled workers will not work for a contractor who neglects the safety rules.

See that your workers wear appropriate clothing for the job at hand, hard hats, work gloves, heavy shoes, and safety glasses as needed. If you have an accident-prone worker in your crew, talk to him and try to convince him that he should correct his ways. If he does not respond, it is recommended that he be transferred to a less hazardous job or be discharged.

Many contractors require accident reports, which should be filled out completely, and it is recommended that you keep your own diary on accidents, recording what, when, where, and why, just in case you are called to testify in court.

Causes of accidents include crane or backhoe booms touching overhead power cables, steel beams or other items slipping loose while being hoisted, material being placed too close to operating machinery, continuing to use cable or rope after some strands have broken, loose trench banks, improper sheeting, careless use of explosives, and pieces of lumber with nails protruding.

20.2. Sandblasting

Sandblasting is used to clean the brick or stone facing on old buildings, to knock the rust and scales off of metal, and to remove laitance from concrete.

The sand is forced at high pressure through a special type of hose and nozzle and is sprayed against the surface to clean it in much the same way as a stream of water would. The operation creates a great deal of dust, and the sand bounces off the surface, so the workers should wear a respirator and safety goggles.

The blasting can damage the mortar joints of old masonry so that the joints will require repointing.

Sandblasting is usually done by a subcontractor who has the special equipment and trained operators required.

20.3. Sand Hogs

People who work under air pressure in pneumatic caissons and tunnels are called sand hogs. The worker, on her knees, will straddle a discharge hose and scoop the sand and mud into the end of the hose with a wooden paddle. The high air pressure inside will force the sand and mud out through the hose. The scooping is similar to the rooting of a hog in the soil.

Sand hogging is a strenuous occupation and can be dangerous. Those who undertake it should be in very good health. The greatest danger is from entering or leaving the chamber too fast, especially leaving. Too fast a decompression can cause the "bends" or caisson disease, which affects the nervous system, causing extreme pain.

If you are interested in this kind of work, you should serve an apprenticeship under highly skilled supervision. People can suffer the bends in accidents in which the pressure in a large caisson drops for a few minutes and then slams on again. The pain can last for several years.

20.4. Sand Piers

When a highway embankment is to be placed in an area of high groundwater or in a swamp, sand piers are used to speed up the settlement of the ground so that it will be stable after the pavement has been constructed. Figure 20.1 shows the method in general. Sometimes the equipment for drilling the holes will be a screw device within a steel casing. After the hole is drilled to the proper depth, the screw and casing are removed while compressed air is applied to keep the hole open. Other methods withdraw the screw and then pour in the sand as the casing is being raised. The sand should be coarse and uncompacted to allow free movement of the groundwater. The settlement gauges should be set as shown, with a bottom plate about 18 in. square. These should be placed not over 100 ft apart, with the vertical rod protruding as shown. The drain pipes should be 4 or 6 in. in diameter and connect to a box set over the pier so as to prevent embankment material from plugging the sand. The drain pipes should carry the water well away

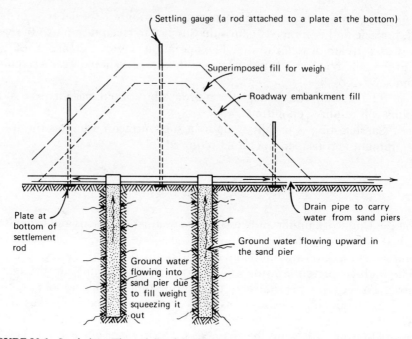

FIGURE 20.1. Sand piers. The weight of the fill compresses the soil and squeezes the ground-water out into the sand piers. When the flow stops and the settling is complete, the superimposed earth is removed. The settling rods are checked at regular intervals by the surveyors, and the movement is reported to the soils engineer.

from the fill to a point where it cannot filter back into the area. The amount of superimposed fill is usually determined by experience or trial and error. If the amount deposited does not compress the area enough, additional fill is placed.

When the settlement is completed, the drain pipes under the fill should be plugged to prevent surface water from filtering back into the ground.

20.5. Sanitary Landfills

A construction foreman would rarely be involved with a sanitary landfill; however, it might be good to know something about them.

A sanitary landfill is regulated garbage dump. The garbage is dumped in layers and compacted by a dozer leveling off the area. When the layer has reached a specified depth, a layer of earth is placed over it, and garbage is again deposited. Such a fill can reach a depth of several feet. After such a fill has reached its maximum depth, it may sit for several years before structures are placed upon it. Such structures will be founded on piling, and problems may arise during the driving (see Section 16.3).

20.6. Scaffolding

Scaffolding is the series of platforms the workers stand on while working above ground. (See Figures 20.2–20.4.) Most states have strict laws governing the design and materials used for scaffolds. On most construction projects, the scaffolding will be welded steel pipe erected in sections, one section fitting on top of another and being diagonally braced. The bottom section will sit on an adjustable base, which operates as a screwjack to keep the scaffolding vertical and stable. Scaffolding used to be, and in some cases still is, constructed of timber, but most contractors use the commercial type, as it is much faster to erect and dismantle.

Wood planks will be set on the scaffolding frame to form work platforms. The planks should overlap the bearing by about a foot so that if they are moved slightly endwise, they will still stay in place. It is recommended, and some states require it, that all scaffolding over two sections high be tied to the structure if possible or be secured by guy lines.

FIGURE 20.2. Steel-pipe scaffolding on the end of a new motel structure. Each section, when assembled, is about 6 ft high, 6 ft long, and 4 ft wide. The sections fit together like an erector set. They can be stacked as high as necessary. Wooden planks are placed on top of each section to provide a platform for the workers. Screw jacks on the bottom of each vertical pipe keep the system vertical. (See Figure 20.4.) Some contractors tie the scaffolding to the wall after the mortar has set to provide added stability.

FIGURE 20.3. Scaffolding for masons erecting a cement-block and brick-faced wall. Note the forklift being used to raise materials for the masons on the platform.

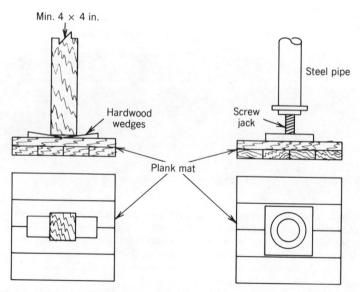

FIGURE 20.4. Shows a footing mat to support the columns of falsework or scaffolding.

If your crew is working on scaffolding, they should be careful not to drop anything that might injure persons below, or if the crew is working near scaffolding, they should be alert to any dropped object and have their hard hat on.

Special care must be taken in winter weather to keep ice and snow off the planks. Sand spread over the planks will improve the footing.

When the planks overlap on a long scaffold, a cheap safety device is to drill a hole at the center of the overlap and insert a ½-in. bolt to keep the plank in line.

20.7. Scarifying

Scarifying is used to loosen the soil or old roadway surface to provide a better bond when material is to be added.

The scarifying will be done by heavy cast-iron or steel teeth mounted on a roller or road grader. The teeth are forced down into the soil several inches as the equipment is moved forward, breaking up the surface into small pieces. When new material is added and compacted, the old and new will bond together. It may take several passes to break the surface into small pieces. Sometimes a heavy disc harrow will be used to break the soil down even further.

20.8. Screeds

The tool used to level off the surface of freshly placed concrete is called a screed. It may be a small hand-held piece of board for small jobs like a sidewalk, it may be a metal plate with a long handle for floor or highway slabs, or it can be a vibrating machine running on the side forms of a highway or other slab. The purpose is always the same: to level off and smooth out the surface of the concrete slab.

It is best to have a worker in front of the screed with a square-point shovel to knock down the high points and fill in the low ones so that the screed is touching concrete for its full length as it moves forward. With the forward movement, the screed should be moved back and forth to shave the concrete down to the proper level. When finished, the surface should be smooth, with all the low places and pinholes eliminated and with a true even finish.

The tops of the side forms will be set at the correct elevation so that as the screed moves back and forth over them, the surface of the concrete will be at the proper elevation. For a large surface area like a floor slab, there may be intermediate rails to keep the screed at the proper elevation.

20.9. Sealants

A liquid sealant is sprayed or painted on a masonry surface to seal the pores and make it waterproof. Other types are forced under pressure into small cracks in masonry, concrete, or other surfaces to seal out moisture.

The sealant may be a chemical, epoxy, paint, or other material.

20.10. Servicing (Equipment)

Most contractors will have a mechanic service all equipment on a regular schedule. It is a proven fact that preventive maintenance saves many dollars. However, many contractors neglect this servicing, especially minor adjustments, with the result that some machines do not operate at top efficiency, and this can affect the production of your crew. If you get in this position, it is recommended that you try to arrange with your superior to give you the means for doing the servicing yourself. Keeping "your" crew working at top efficiency will be noticed at the front office.

20.11. Settling Tanks

The suspended solids in sewage or industrial waste are removed by a slow flow through a settling tank. The tanks may be rectangular or circular. The flow will be over V-notch weirs, which spread out the flow rather than having it concentrated at one point. The velocity of flow through the tank (retention time) will be governed by the type and concentration of solids.

There will be boards (sweeps) along the bottom of the tank to carry the solids to a pit from which the solids (sludge) will be pumped to a digester where bacteria will reduce the volume.

There will be small rails on the bottom of the tank over which the boards will slide, and the boards will have metal shoes to ride on the rails and reduce wear. The boards will be attached to an endless chain, which will cause them to slide along the bottom, carrying the solids to the pit. Then the boards will rise above the water surface and be conveyed back to the inflow end of the tank where they will be again conveyed to the bottom of the tank and repeat the cycle. On circular tanks, the boards or sweeps do not rise from the rails, but sweep in a circular motion, which conveys the solids to the pit at the center of the tank.

The boards move very slowly, only a few feet a minute, and must be set to move freely so as not to bind.

The rectangular tank floor will be flat or slope very little toward the pit. The circular tank floor will have much more slope toward the pit at the center.

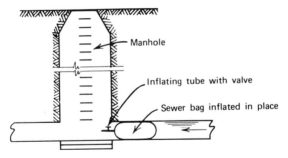

FIGURE 20.5. Sewer bag. The sewer bag is inserted into the sewer pipe and inflated as indicated.

20.12. Sewer Bags

A sewer bag is used to temporarily close off a section of a sewer for testing or repairs. The bag is inserted in the sewer line as shown in Figure 20.5. With a short hose and a value on the end, the bag is inflated to a pressure sufficient to hold against the water. The amount of pressure will depend on the head of water to be resisted. When testing for exfiltration, the head of water may be too much for the bag, and it will be necessary to add timber bracing.

20.13. Sewers

Sewers can be only sanitary, only storm, or a combination of the two. It is best to keep the two separate, but sometimes that is not economically possible. When the two are combined, the low sanitary flow will go directly into the treatment plant, and when storm water is added, a jump weir will cause the flow to bypass the plant. (See Figure 24.18.)

Sewer pipe is made of vitreous clay, reinforced or plain concrete, asbestos–cement, coated steel, cast iron, or plastic. The clay, plain concrete, and asbestos pipes are seldom over 24 in. in diameter. Most large-diameter pipe will be reinforced concrete.

As most sewers flow by gravity, the slope of the pipe is very important. (See Section 2.10, Figure 2.4.) The surveyor's stakes will indicate the alignment and slope (grade). It is the foreman's job to set the batter boards on most projects, although the carpenters do so on many. You must know what each stake represents for both horizontal and vertical measurement. The batter boards must be set sturdily in place and kept undisturbed until construction in that area is complete. The line between batter boards must be kept taut; otherwise, there will be a dip in the sewer line. It is recommended that you constantly check the pipe to keep it on grade, as it is much easier to correct a misalignment before you get too far ahead or the pipe is backfilled. (See Sections 3.14, 4.32, 13.5, 13.18, and 17.17.)

The pipe will come with the gaskets in place or loose for installation on the job. In either case, they must be clean and well lubricated with grease supplied with the pipe. The bells should be laid upstream, and each section should be rammed tight into the previous section. An easy way to be sure that the pipe is up tight is to use a crowbar and a block of wood as shown in Figure 2.12.

The clay pipe, particularly the smaller-diameter pipe, may have a slight bend, and if the inspector will permit it, may be laid with the bend horizontal. The sewage will flow around the bend without difficulty and still maintain the grade.

The plain concrete pipe is used mostly for drainage lines. This pipe requires reasonably careful handling, as it is not too difficult to damage, particularly the spigot end.

On the other hand, the reinforced concrete pipe is much tougher, particularly when it has a steel core and bell and spigot. However, it is always wise to handle all pipe with care.

The asbestos–cement pipe (A/C) also requires careful handling. It is easy to handle because of its light weight, so it is quite often placed in the trench by hand. But care must be taken not to damage the spigot end as it is much thinner than the rest of the pipe. If the spigot is damaged, the gasket may not fit properly.

The coated steel pipe is lightweight and comes in long lengths (12–20 ft). It is usually corrugated, with the connecting collars fitting into the corrugations. The ends of the pipe must be up tight so that the collar will fit tightly. If the coating is bituminous, it may get sticky in the hot weather. if the coating is cement, a sharp blow may break it loose.

Cast-iron pipe is the heaviest and strongest, but it can be damaged by being handled too roughly. Cracks in this pipe can be hard to detect and may show up and cause problems only if it is under pressure. Fine cracks can be detected by "ringing" the pipe, that is, suspending the section with a rope sling and striking it with a hammer. If the pipe is sound, it will ring like a bell, but if cracked, there will be a dull thud.

Plastic pipe is, size for size, the lightest weight and has a thin wall. It comes in long lengths and is usually connected by a collar. In making the connection, the ends of the sections are cleaned and treated with a compound before the collar is slipped on. The compound will weld the pipe together. Like the A/C pipe, the plastic pipe will resist most acids better than the other pipes. Also like the A/C pipe, the plastic needs a well-compacted bedding.

All bell and spigot pipe, regardless of type, should have a bell hole dug so the pipe will bear on its full length instead of just the bell. (See Section 3.14.)

FIGURE 20.6. Sewer force main. A 36-in. force main is set on a reinforced concrete mat and cradle. High-early-strength cement was used so that the pipe could be installed 24 h after the mat was poured.

FIGURE 20.7. Shows a trench containing a large force-main and a smaller gravity sewer. Note the sheeting is supported by soldier beams which eliminates cross-struts. The cross timbers in the background are to support the concrete shute.

Uncoated corrugated steel pipe is often used for storm sewers, and as storm water carries considerable silt, it is important that the ends be up tight and the collars be tightly fitted. Sections of 30-in. sewer have been destroyed because the joints were not up tight, and the silt and sand in the fast-moving water ground up the gasket, loosening the collar and allowing the water to leak out and undermine the sewer.

Following are several photos of various sewer projects.

Figure 20.6   Note that the pipe in the background has been set on the mat, but the cradle has not been poured. There is a length of concrete mat poured just in front of the pipe. Note the reinforcing bars in front of the mat. Enough mat was poured each day to carry two lengths of pipe.

Figure 20.7   The figure shows a large reinforced-concrete force main with a small cast-iron gravity sewer alongside it on the same mat. Note that the soldier beams and timber sheeting eliminate the cross struts.

Figure 20.8   The figure shows the same trench before the mat was poured. The timbers across the trench are to hold the chute for placing the crushed stone lining and the concrete for the mat and cradle.

FIGURE 20.8. Shows the same trench as in Figure 20.7 but before the concrete mat and pipes have been installed. The bottom of the trench has been lined with crused stone ready for the concrete. Note the chute for the concrete.

FIGURE 20.9. A 24-in. reinforced-concrete sewer force main on a creosoted-timber mat in silty clay soil. A concrete cradle will be installed under the pipe.

FIGURE 20.10. A 72-in. reinforced-concrete storm sewer being installed on a reinforced mat. The pipe is set on concrete blocks 8 in. above the mat, and a concrete cradle will be placed to bed the pipe. Note the cable sling by which the crane sets the pipe.

FIGURE 20.11. A 72-in. reinforced-concrete storm sewer being encased in concrete in an area under an airport runway. The sheeting and bracing will be left in place. Note the vertical struts to keep the pipe from floating during the placing of the concrete. The concrete is brought up on both sides of the pipe simultaneously to prevent the pipe from being pushed off line. Vibrators are used to force the concrete firmly and fully under and around the pipe.

Figure 20.9   The figure shows a 24-in. reinforced-concrete force main on a creosoted-timber mat on silty clay soil. A concrete cradle is to be installed under the pipe. Note the steep trench banks without sheeting as the area is stiff, dry clay.

Figure 20.10   The figure shows a 72-in. sewer being placed on a concrete mat as noted. The sewer provides drainage for a large airport.

Figure 20.11   The figure shows the same sewer in another area where tight sheeting was required. The pipe was fully encased in concrete as it was under the runway.

20.14. Shear Breaks

Shear breaks occur when underground pipes, particularly nonmetal pipes, are not properly bedded. Uneven compacting of the subgrade or lining will result in uneven support when the backfill is placed over the pipe. This uneven pressure can cause shear breaks. The break may collapse the pipe, but more often a creak completely around the pipe will result, and the section with the least support will settle, allowing infiltration if there is groundwater and exfiltration if there is none. (See Section 10.3.) If there is a crack but no misalignment, any leaking can be stopped by chemicals. (See Figure 21.22.)

20.15. Shear Pins

Equipment that can become bound or stuck during operation should have a shear pin on the drive shaft of the power unit to protect it from damage. The pin will be of softer metal than the shaft so that when any part of the equipment becomes stuck the pin will shear. A good example of this is in the machinery that moves the sweeps in a settling tank. If the boards become bound in any manner, the shear pin on the power shaft will shear, allowing the power unit to run free and not damage the unit or break the chain on the sweeps.

20.16. Sheeting and Bracing

The amount of sheeting and bracing required will depend on conditions. The required amount can vary from a 2 in. plank set vertically every few (5 or 6) feet and held in place by a timber strut or trench jack to full tight sheeting. (See Figures 20.12–20.14 and Tables 20.1 and 20.2.) Shale, rock, or hard-packed earth will require no sheeting, while loose, damp, conditions will require the tight sheeting. Sheeting and bracing slows the progress of the work and costs money, and most contractors are reluctant to use it unless it is absolutely necessary. They may decline to use sheeting unless ordered to do so by the project, or OSHA, inspector. You must know the contractor's policy on these matters. If he is reluctant to use sheeting and you believe the conditions warrant it, you should warn him and let him overrule you if he wants to. Thus, you are protected if a problem arises.

The timber used for sheeting and bracing should be good quality, at least 1000 lb fiber stress. Some contractors use cheap lumber for sheeting which has a low fiber stress and is brittle due to being dried out. If that is the case on your project, you should warn the workers to be alert for snapping or popping sounds, which usually precede a failure of low-grade timber. Low-grade lumber will bend very little and snap off quickly when failing. High-grade lumber will give considerably before failing and will usually break slowly, giving the workers a chance to move out of the area.

Some specifications require that sheeting be placed and maintained in contact with the earth as it is placed. Under such conditions, the planks should be sharpened at the bottom, with the point on the outside. Sometimes, on sewers which must be maintained on grade, the sheeting will be driven well below the subgrade, and when it is pulled, a void will be left, which can let the subgrade move laterally and cause the sewer pipe to settle. These voids should be rammed full as the individual sheeting planks are withdrawn. (See Figure 23.2.)

When sheeting and bracing are to be removed, the trench should be backfilled up to the bottom of the first set of walers; these walers should be removed and the backfilling should be continued to the next set of walers,

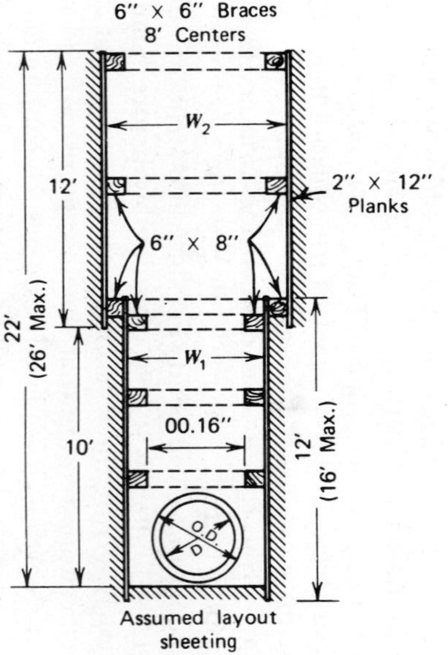

Sheeting Quantities

Depth trench ft.	Sheeting Length ft. 2"×12'	B.F. Lin.ft.	Bracing No. Braces Req'd 6"×8"	B.F. Lin.ft.	No. Braces Req'd 6"×6"	Bd.ft./sq.ft. R.C.P.	Bd.ft./sq.ft. V.P.	Total sheeting R.C.P.	Total sheeting V.P.
6	7	28	4	16	2	6	4	50	48
8	9	36	4	16	2	6	4	58	56
10	11	44	4	16	2	6	4	66	64
12	13	52	6	24	3	9	6	85	82
14	16	64	10	40	5	15	10	119	114
16	18	72	10	40	5	15	10	127	122
18	20	80	12	48	6	18	11	146	139
20	22	88	12	48	6	18	11	154	147
22	24	96	12	48	6	18	11	162	155
24	26	104	12	48	6		11	170	163
26	28	112	12	48	6	18	11	178	171

R.C.P. = reinforced concrete pipe
V.P. = vitrified pipe

FIGURE 20.12. Sheeting and bracing quantities. Recommended sizes and number of bracing timbers for the various depths listed.

FIGURE 20.13. A 24-in. reinforced-concrete trunk sewer installed within a trench protected by sheeting and bracing. The sheeting in the foreground will be driven to an even depth, the same as that in the background. The sheeting and bracing will be left in place because of unstable soil conditions in a built-up area. The trench will be backfilled with compacted run-of-bank gravel. The pipe is on a reinforced concrete mat and cradle.

FIGURE 20.14. A double set of sheeting and bracing in a deep trench. The lower set of walers (timbers) is set at the top of the bottom set of sheeting. The top set of sheeting is set just outside the lower set and overlaps about a foot. The workers are preparing to set the upper set of walers.

Table 20.1. Minimum Sizes of Bracing and Sheet Piling for Narrow Trenches[a]

Depth of Trench (ft)	Sheet Piling		Stringers		Cross Bracing	
	Size (in.)	Horizontal Spacing	Size (in.)	Vertical Spacing	Size (in.)	Horizontal Spacing
(a) Hard and solid soil						
5 to 10	2 × 6	8 ft	2 × 6	4 ft	2 × 6	5 ft
10 to 15	2 × 6	2 ft	2 × 6	4 ft	2 × 6	4 ft
More than 15	2 × 6	tight	4 × 10	4 ft	4 × 10	6 ft
(b) Soil likely to crack or crumble						
5 to 10	2 × 6	3 ft	2 × 6	5 ft	2 × 6	5 ft
10 to 15	2 × 6	2 ft	2 × 6	4 ft	2 × 6	4 ft
More than 15	2 × 6	tight	4 × 10	4 ft	4 × 10	6 ft
(c) Soft, sandy filled-in loose soil						
5 to 10	2 × 6	tight	4 × 6	6 ft	4 × 6	6 ft
10 to 15	2 × 6	tight	4 × 6	5 ft	4 × 6	6 ft
More than 15	2 × 6	tight	4 × 12	4 ft	4 × 12	6 ft
(d) Where hydrostatic pressure exists						
To 10	2 × 6	tight	6 × 8	4 ft	6 × 8	6 ft
More than 10	3 × 6	tight	6 × 10	4 ft	6 × 10	6 ft

[a] Not more than 4 ft in width.

and so on. One of the easiest ways to remove sheeting is by use of a U-shaped tool. (See Figure 20.15.) This tool, attached to the cable of a crane, will pull the sheeting planks quickly and without breaking them.

Some contractors use 4 × 8 ft sheets of ¾ in. plywood braced with timber or trench jacks. These are a little unwieldy, however, and the planks are preferable.

For difficult areas or deep excavations, steel sheet piling is used (see Section 20.17).

When tight sheet piling is used, the cross struts can interfere with the work, so it is advisable to have heavy walers to keep the span between the struts as long as possible.

When calculating the desired width of a trench with full sheeting and bracing, the distance from outside to outside of the sheeting must be such as to allow the required working space between the walers.

When starting a sheeting and bracing installation, the framework of walers and struts should be set first, and the sheeting should be driven outside the walers. The sheeting is usually driven by an air-operated jack-hammer which has a special head to fit over the end of the sheeting planks.

Table 20.2. Minimum Sizes of Bracing and Sheet Piling for Wide Trenches[a]

Depth of Trench (ft)	Sheet Piling		Stringers		Cross Bracing	
	Size (in.)	Horizontal Spacing	Size (in.)	Vertical Spacing	Size (in.)	Horizontal Spacing
(a) Hard and solid soil						
5 to 10	2 × 6	6 ft	4 × 6	4 ft	4 × 6	6 ft
10 to 20	2 × 6	tight	6 × 6	4 ft	6 × 6	6 ft
More than 20	2 × 6	tight	6 × 8	4 ft	6 × 8	6 ft
(b) Soil likely to crack or crumble						
5 to 10	2 × 6	3 ft	4 × 6	4 ft	4 × 6	6 ft
10 to 20	2 × 6	tight	6 × 6	4 ft	6 × 6	6 ft
More than 20	2 × 6	tight	6 × 8	4 ft	6 × 8	6 ft
(c) Soft, sandy filled-in loose soil						
5 to 10	2 × 6	tight	4 × 6	4 ft	4 × 6	6 ft
10 to 20	2 × 6	tight	6 × 6	4 ft	6 × 6	6 ft
More than 20	2 × 6	tight	6 × 8	4 ft	6 × 8	6 ft
(d) Where hydrostatic pressure exists						
To 10	2 × 6	tight	6 × 8	4 ft	6 × 8	6 ft
More than 10	3 × 6	tight	6 × 10	4 ft	6 × 10	6 ft

[a] Not more than 8 ft in width.

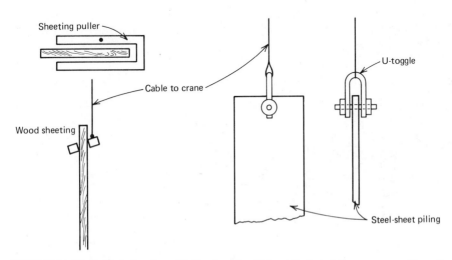

FIGURE 20.15. Pulling sheeting. All sheeting should be pulled straight up, as any side pull will increase the friction and make it more difficult to pull.

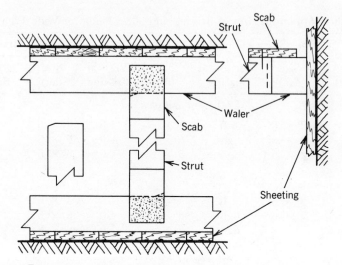

FIGURE 20.16. Shows a method of installing sheeting and bracing. Workers will hold the walers tight against the sheeting while others install the struts. The struts must be driven tight to keep the walers in place. Note a bevel has been cut on opposite corners of the strut to facilitate driving it tight against the walers.

Some contractors excavate several feet with a backhoe before starting the sheeting and then continue excavating with a crane and bucket.

When the soil is reasonably firm, the excavation can be carried several feet (3 or 4) below the bottom of the sheeting before it is driven down. Care must be taken to see that the sheeting plank stays tight against the earth and remains vertical. To keep the plank tight against the earth, excavate

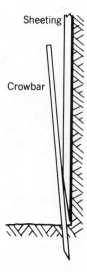

FIGURE 20.17. Shows a method of keeping the sheeting tight against the earth behind.

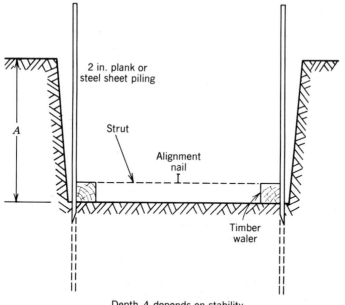

2 in. plank or
steel sheet piling

Strut

Alignment
nail

A

Timber
waler

Depth A depends on stability
of the soil. It is usually 2 or 3 ft.

FIGURE 20.18.

FIGURE 20.19. What sometimes happens when a contractor ignores advice and tries to get by with slipshod methods. In this case, passing through water-bearing soil, well points were not used. As silt-bearing water was pumped from the trench, a blow-in occurred, causing the sheeting to settle as shown. Note the pump set on the walers in the background. After installing well points and resetting the walers, the work resumed.

slightly less than required and let the point of the plank shave off the remaining earth as it descends.

A sheeting plank can be kept tight against the adjacent plank while being driven by pressure from a crowbar or shovel held against the other edge.

One easy way to remove wood sheeting is by use of a U-clamp or, for steel sheeting, a U-toggle. (See Figure 20.15.) If the sheeting is to be used again, it should be cleaned while the earth on it is still damp. Wood sheeting should be piled in such a manner as to prevent warping, and steel sheets should be given a coat of light oil.

20.17. Sheet Piling (Steel)

Steel sheet piling comes in several shapes and weights. (See Figure 20.32.) It is used to retain the sides of earth excavations, to construct cofferdams, retaining walls, piers, and docks, and any place the earth must be retained.

The sheets have a connecting groove, similar to a ball and socket joint, along each edge so that the sheets interlock to form a solid wall. These grooves should be greased before being installed and cleaned as soon as they are removed, while the earth in the groove is still moist.

The sheets are driven by a special steam- or air-driven hammer that fits over the sheet. There is also a vibrating hammer that can be used. Figure 20.15 shows one method of removing the sheets.

On some projects, the sheet piling will be driven to the full depth before excavation starts, and the walers and struts will be installed as the excavation progresses.

20.18. Sheetrock (Wallboard)

Sheetrock is wallboard made of plaster covered with a moisture-resistant paper. It replaced the old wood lath and plaster. When the wallboard is to be covered with plaster, it will come in a 16 × 48 in. size and when used as "drywall" it will come in 4 × 8 ft sheets. The drywall joints will be taped and plastered and the area will be sanded after the plaster has set to present a smooth surface ready for paint or wallpaper. The sheets will be attached to the studs by special nails, and with drywall, the nails will be set slightly and the hole will be plastered and sanded.

Sheetrock stored before use should be kept dry, and when handled, care must be taken not to damage the edges.

20.19. Shields (Excavation)

Figures 20.20–20.22 will show you what a shield is and how it is used. The shield will be made of heavy steel sheets with the width at the top a few

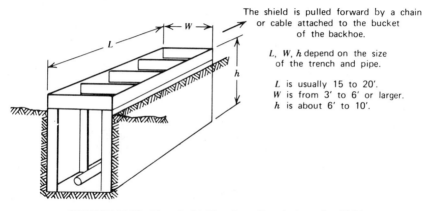

The shield is pulled forward by a chain or cable attached to the bucket of the backhoe.

L, W, h depend on the size of the trench and pipe.

L is usually 15 to 20'.
W is from 3' to 6' or larger.
h is about 6' to 10'.

FIGURE 20.20. Trench shield or box. One design of a shield.

FIGURE 20.21. A metal shield being pulled forward by the backhoe. Note that the trench behind the shield has been partially backfilled to prevent the banks from collapsing.

FIGURE 20.22. Trench shield. Trench shields in use in unstable earth. Note that the backhoe is on timber mats.

inches greater than at the bottom to reduce the friction with the soil as it is raised or moved. The workers within the shield are protected as they prepare the subgrade, place lining, and install the pipe. When the work in the shield is complete, the backhoe will excavate the trench and pull the shield ahead far enough to allow the laying of another length of pipe.

The bottom of the shield should be no lower than the invert of the pipe. If lining is used, it should be below the bottom of the shield. If well compacted lining is placed between the sides of the shield, it may grip the pipe and pull it apart as the shield moves forward.

The shield should be raised a few inches straight up to break the suction before being pulled forward. If the work is in an area of damp clay, the suction on the sides of the shield can be considerable.

20.20. Shimming

There are various kinds of shims. Small steel wedges are used under column-bearing plates to hold them level until the grout can be placed and set. Thin metal sheets, usually copper, are used between the halves of a bearing housing when it is too tight. Thin sheets of metal are used under equipment bases to make the units level when they are bolted down.

20.21. Shooting (Blasting)

When blasting is to be undertaken, it is customary to say "we will shoot it," rather than "we will dynamite it" or "we will blast it."

20.22. Shop Drawings

The engineer's plans on which a contractor bids do not always show various parts of the structure and equipment in detail, as the contractor is allowed to choose the supplier, and the choice is not known at the time the plans are prepared. When the contractor awards an order to his supplier, the supplier will prepare and submit detailed drawings of what she will furnish. These drawings are called "shop drawings" because they are made by the shop that furnishes the material. These drawings will have all the details necessary to install the material or equipment being furnished. If you are given shop drawings to follow on your assignment, keep them clean and safe and return them to your superior, so that in case of an argument over how some unit should be erected, the drawings will be available to check.

20.23. Shutdowns

A shutdown of the work may be caused by adverse weather, an accident, a change in design, a delay in the delivery of some material or equipment, or OSHA, due to some violation of their regulations. If your part of the work is to be shut down, follow the instructions of your superior as to what to do before leaving or delaying the work. You should get details on how long the shutdown will last, if known, and on what precautions you must take to protect the work already done so you can advise your crew members.

20.24. Sight Distances

The sight distance is the distance at which an operator of a vehicle can first see an approaching vehicle on a horizontal or vertical curve. (See Figure

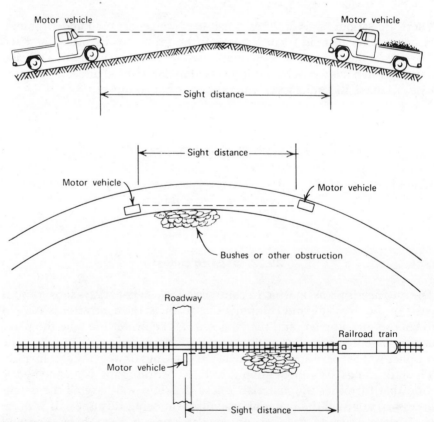

FIGURE 20.23. Sight distances. Sight distance is the distance at which a vehicle operator can first see an approaching vehicle or other obstacle over a vertical or around a horizontal curve. The law specifies what the sight distance must be for railroads and highways.

20.23.) If you are involved in clearing for sight distance, you should arrange to have your eye level at the elevation where the operator would be when in his vehicle, and you should have targets indicating the top and width of the largest expected approaching vehicle. All brush, trees, and other obstacles should be removed so that the sight is clear, both horizontally and vertically.

20.25. Sika

Sika is a brand name for a chemical used with cement mortar to seal leaks in masonry and concrete structures. The instructions with the container must be carefully followed, especially when using the quick-set type, which can burn the skin. There are other such chemicals under other brand names, all of which should be used only as per the manufacturer's instructions.

20.26. Siphons

A siphon is a device which uses the atmospheric pressure to carry water, or other liquid, from one elevation, over an obstacle, to a lower elevation. Figure 5.12 shows that if a pipe or tube is filled with water and the upper end is submerged, the water flowing out of the lower end will cause a vacuum, which will pick up the water in the higher level, and the flow will continue until the vacuum is broken.

Siphoning is an economical method of dewatering a large hole or excavation if there is a lower level to which the pipe can be carried for discharge and time is not important. The siphon is, of course, much slower than forced pumping.

The siphon is filled by closing the valve at the discharge end, raising the pipe as shown at *A* in Figure 5.12, putting a temporary plug in the end of the pipe (a rag or wooden plug will work) and filling the pipe with water. The pipe is than swung down into the water as shown at *B*, removing the plug as the end of the pipe enters the water and simultaneously opening the value. The water will flow as long as the upper end is below the water surface. The joints in the pipe must be tight. Slight leaks can be sealed with a coating of clay or thick mud.

20.27. Slabs on Grade

"Slabs on grade" means concrete slabs poured on the ground, such as for the ground floor of a structure, for a highway pavement, or for a driveway or parking area. The slab may or may not be reinforced. The reinforcing may be wire mesh or steel bars. The design should be limited to 10 ft in any one direction for unreinforced slabs and to 50 ft for reinforced slabs, depending on the kind of reinforcement used. You, of course, will be limited to what is shown on your plans.

Before a slab is poured, the subgrade must be correct. It must be to the proper grade, with all mud and water removed. In cold weather, frost must be removed. If the frost is only on the surface and has not expanded to the ground below, it can be scraped off with a hand shovel or grader blade. If the earth has expanded, the frost must be removed for the full depth and the earth must be compacted. Frost can be removed by applying heat, or better yet, by preventing frost by covering the area with 12 in. of straw and canvas or with plastic sheets.

The typical highway slab will be 8–10 ft in width and 50 ft long, although some are now poured several hundred feet long with special heavy reinforcement. Floor slabs will usually be limited to 20 ft in any one direction and have light-gauge mesh reinforcement. Such floors will be poured in alternate sections. Most slabs will have a broom or burlap finish to provide traction, and the edges will be tooled with an "edger" to eliminate the sharp corners. Floor slabs that are to have a covering applied will be finished smooth.

Slabs exposed to the elements must be protected from frost in cold weather and too-rapid drying out in summer. A cover of 12 in. of loose straw and canvas or plastic to hold it in place will protect against all but extreme cold. In summer, the slab can be sealed with bitumen or other sealant to keep it from drying too fast, or wet burlap can be used. Some contractors use 3 or 4 in. of damp earth and blade it off when the concrete has fully set.

20.28. Sleepers

When a hardwood floor is placed over a concrete slab, the flooring strips will be held in place by wood or metal inserts cast in the surface of the concrete. These inserts are called "sleepers" and are placed about 16 in. apart at right angles to the flooring strips. No. 6 or No. 8 finish nails will be used with the wood inserts, and special nails or screws will be used with the metal. The nails will be driven into the edges of the strips so as to be hidden in the joint. The screws will be countersunk on the surface. Sleepers are similar to furring, except that the furring is attached to the surface rather than embedded.

Some specifications will require that the concrete slab be covered with a seal of hot tar or other sealant to keep the moisture from rising into and warping the flooring.

20.29. Sleeves

On civil engineering projects, "sleeve" refers to the threaded tube used to connect two sections of metal pipe. It may also refer to a short section of pipe cast into a concrete or masonry wall through which smaller pipes are installed (see Section 24.3).

20.30. Slip Forms

Slip forms slip or slide either horizontally or vertically as the concrete is poured and retain it until it has set enough to carry the load. For horizontal pours on the ground, the forms have to retain the concrete only until it has set enough to stay in place, such as for slabs or curbs. Vertical pours for walls or chimneys must have forms that will retain the concrete until it is set enough to support the concrete, the weight of the forms, and the equipment that makes the forms move. The vertical pours will use high-early-strength cement.

This work is handled by specialty contractors who have the equipment and trained crews.

20.31. Sludge

The solids or thick residue that remain after treatment of sewage or industrial waste are called sludge. It will settle to the bottom of the tank and as in the settling tank, be swept to a pit from which it will be pumped to disposal. The disposal may be by drying bed or by vacuum filter, after which the sludge may be used as fertilizer, burned, or buried.

20.32. Slump Tests

The consistency of concrete is usually checked by a slump test. The job inspector or laboratory personnel will take a sample of the fresh concrete as it flows from the truck or mixer and place it in a metal cone, as shown in Figures 20.24 and 20.25. If your crew is handling the concrete you should cooperate in the taking of samples.

20.33. Slurry

Slurry is a mixture of water and clay or Bentonite of the consistency of thick soup. It is used when trenching in unstable soil to hold the banks from collapsing. As the trench is dug, the slurry is pumped in, and the machine works through the slurry to excavate to the required depth. The pressure of the slurry against the banks keeps them from collapsing. When the trench is being backfilled, the slurry will be pumped off. Pipes can be laid in the slurry in the same way as underwater. (See Section 17.17) (See Figure 20.26.)

FIGURE 20.24. Slump test. An inspector making a slump test and casting four test cylinders. Note that the slump test cone has been placed bottom end up and that the ramming rod has been placed on it, extending over the slumped concrete. The distance from the underside of the rod to the top of the slumped concrete is the recorded slump. The test cylinders are allowed to set for 24 h. Then they are taken to the laboratory for controlled temperature and humidity storage until broken at 7, 14, and 28 days.

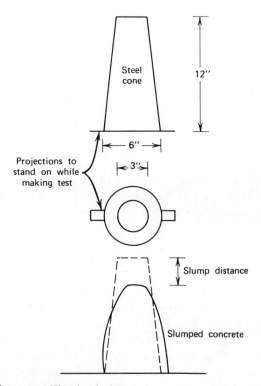

FIGURE 20.25. Slump test. The sketch shows a test cone and the way the concrete slumps in relation to the cone height.

328

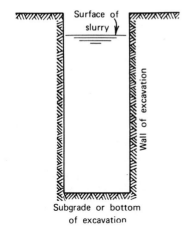

Surface of
slurry

Wall of excavation

Subgrade or bottom
of excavation

FIGURE 20.26. Slurry used in trench excavation. The backhoe or other excavator operates in the trench through the slurry. The excavated material is raised slowly to allow the slurry to drain off. The slurry, a thick, heavy liquid, holds the walls of the excavation in place. When the excavation is completed, the slurry is pumped out as the backfill is placed.

20.34. Snap-On-Tools

The words "snap-on" are a trade name of one manufacturer of mechanic's tools. A set of socket wrenches or other tools will all fit onto one handle by just "snapping-on" or off as required. This arrangement requires much less storage space and the tools are lighter to carry.

20.35. Sodding

Some projects will require sodding instead of topsoil and seeding, especially when the project is an administration or other building where the public may enter. Usually, the work will be done by a landscaping firm, but your crew might be assigned the job.

The sod should be fresh, with the grass green. The strips should be damp when laid and placed tightly together. If the work is on a slope, the sod should be laid at the base first and should then proceed up the slope so that the strips will stay tightly together. As soon as it is laid, the sod should be sprayed with water and kept damp for several days to give the roots a change to resume growth.

20.36. Soil Augers

Referring to Figure 20.27, this type of power auger is used to obtain soil samples, to check the elevation of the groundwater, and for setting posts. The same type of machine, but with a 12 or 18 in. or larger-diameter auger, is used to excavate for concrete piers or sand piers. The work will usually be done by subcontract, but some large contractors will have their own augers.

FIGURE 20.27. This soil auger is being used to determine the type of soil that will be encountered when trenching for a proposed sewer line. With this information, the contractor can determine the type of equipment to use.

20.37. Soil–Cement

The soil–cement method of construction is an economical way to provide a hard surface for roads, runways, parking areas, and other areas where heavy traffic is not expected. There are several methods of operation, but the preferred one is to scarify the surface to a depth of 4 to 6 in. sufficiently to break the soil up into very small pieces, and to then spread the Portland cement to the amount per square yard specified, sprinkle it lightly unless the soil is already damp enough, and then blade back and forth until the soil with the cement is a uniform color. When a uniform color has been obtained, alternate blading and rolling until a smooth, hard surface is developed. Roadways should have at least a 6-in. crown for quick drainage.

There are several machines on the market for use in this operation, but they basically do the same thing: break up the soil into fine pieces, apply the cement, mix to a constant color, and blade and roll to the required surface.

20.38. Soils

The classification of the soil and its ability to sustain the loads to be placed upon it by the proposed structure is the responsibility of the engineer and his inspector: however, any good foreman should have a general knowledge of the soil. Several authorities have published soil classification data. The Casagrande classification, which describes the nature of the various soils and how they are used, has been one of the most popular publications. The following description of soils is a general one:

1. *Gravel and Sand Mixture, Well Graded from Fine to Coarse.* This mixture drains well and can be well compacted by roller or vibration. As it has no binder, it will loosen up quickly under traffic.
2. *Well-Graded Gravel and Sand with a Mixture of Clay.* Drains very poorly, almost impervious, will compact well by roller, vibrator, and rubber-tired roller. Makes an excellent foundation.
3. *Poorly Graded Gravel and Sand.* Drains well, compacts fairly well by roller. Makes a fairly good foundation, but loosens up under traffic.
4. *Gravel and Sand with a Mixture of Silt and Clay.* Does not drain well, makes a fair foundation, can be well compacted if the moisture content is controlled.
5. *Coarse Sand with Little Fines.* Makes a very good foundation, drains well, can be well compacted by roller or vibration.
6. *Sand with Silt and Clay.* Not too good for a foundation, drains poorly, can be well compacted by rubber-tired roller if the moisture is controlled.
7. *Silts.* Make poor foundations, drain very slowly, can be compacted by rubber-tired roller if the moisture is controlled.
8. *Clay.* Makes a poor foundation, subject to long-time progressive settling. Difficult to compact.
9. *Cemented Gravel.* Makes an excellent foundation, does not need compaction, almost impervious.
10. *Shale.* No drainage, no compaction. Makes an excellent foundation if the loose, soft layer is removed. Should not be exposed to the air too long before the surface will become soft and cracked, so that a layer will have to be removed before the footings are placed.

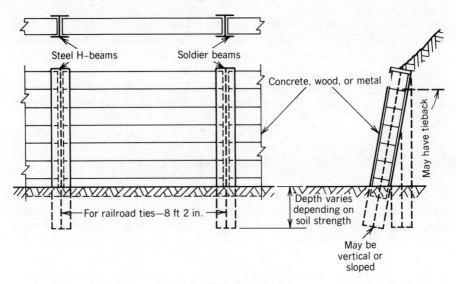

FIGURE 20.28.

20.39. Soldier Beams

Soldier beams are steel H-beams driven into the ground at 8–10 ft intervals with cribbing between them. (See Figure 20.28.) The cribbing may be timber like railroad ties, or of concrete or steel. The beams may be set vertically or at a slight angle. They may cantilever above the ground or have a tieback. A study of the sketch will show the method of installation. If the beams are to cantilever, the ground must be stable to take the pressure.

20.40. Spalling

Spalling is caused when moisture gets into the surface of concrete and freezes. As the moisture freezes, it expands, causing the concrete to flake off, exposing the aggregate and mortar beneath. The exposed concrete is more porous than the finished surface, and moisture can again enter, freeze, and push off more concrete. This cycle can continue until the concrete has been destroyed for a depth of several inches. There have been cases where the spalling on a bridge pier had destroyed the concrete back under the bearing plates and required major repairs to correct.

Spalling can be prevented by a properly rubbed surface that fills all the tiny voids and presents a smooth, tight surface. A further preventive is an application of waterproofing.

When pouring concrete, the area next to the form must be well vibrated and spaded to force the large aggregate at least ½ in. back away from the surface, leaving a layer of well compacted mortar out of which the air bubbles have been forced.

20.41. Special Backfill

Special backfill is used when the excavated material is of poor quality or when a better quality material is required. The specifications will state what is to be considered special backfill, where it is to be used, and the method of payment. The material may be well graded run-of-bank gravel, crusher-run stone, or a specified mixture, or it may be a well graded sand.

Many projects will require special backfill for all or most of the depth of a trench under a highway. It is also used to replace an unstable subgrade. It is usually paid for by the cubic yard in place. A careful record should be kept of where it is used and how much is used.

20.42. Specifications

The specifications, often referred to as the "specs," are the written instructions in the contract documents that specify the kind, quality, and amount of all items in the contract, how the contractor is to be paid and his responsibilities, and when the project is to start and when it must be completed. It will specify the cost to the contractor, if any, that will be assessed if the job runs over the time limit. It will also state the responsibilities of the owner and his engineer. It is a legal document.

20.43. Spillways (Dams)

A spillway allows the high water to spill out of a reservoir or lake without overtopping the dam. Overtopping is especially damaging to earth dams, where the downstream slope can be eroded away. On concrete or masonry dams, the spillway will usually be on the dam itself. (See Figure 5.1.) For earth dams, the spillway will be at one side, with a concrete or masonry channel. (See Figure 19.11.)

Where there is not too much flow and/or the velocity is low, the channel may be lined with riprap. When the volume of water is low, the spillway may be a pipe through or under the dam with or without a valve control.

For construction methods, see Sections 4.35, 6.6, and 19.13.

20.44. Springs (Water)

When excavating, especially in hilly country, it is not unusual to encounter springs, either at the surface or underground. Such springs can destabilize a large area if the water is not drained off. Usually, the engineer or his inspector will indicate the method to be used, which quite often will be a French drain with or without a pipe. (See Section 7.25.) If the spring can be intercepted outside the construction area, it can be rerouted.

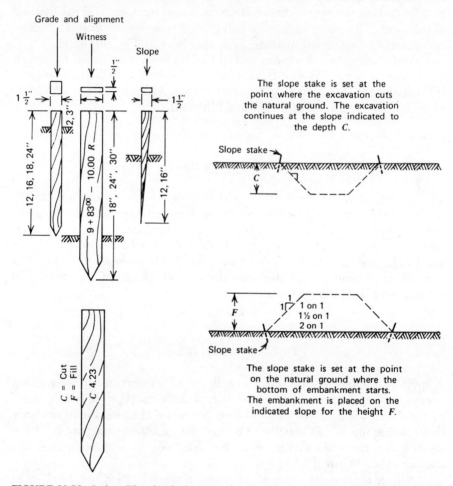

The slope stake is set at the point where the excavation cuts the natural ground. The excavation continues at the slope indicated to the depth C.

The slope stake is set at the point on the natural ground where the bottom of embankment starts. The embankment is placed on the indicated slope for the height F.

FIGURE 20.29. Stakes. The sketch shows grade, alignment, witness, and slope stakes. Slope stakes are set at the point where the excavation will cut natural ground. The excavation will continue at the slope to the depth C. Slope stakes are also set on natural ground at the point where the bottom of the embankment starts. The embankment is placed on the indicated slope for the height F.

20.45. Stakes (Construction)

All kinds of civil engineering construction will be laid out and constructed according to the control stakes; thus, it is important that the foreman know what each stake represents. The surveyor, inspector, or engineer will supply that information, usually on paper. If you do not fully understand the information, ask questions before you start work.

Grade stakes, usually 2 × 2 in. wood, will be driven flush with or slightly above the ground. They will have a tack on the top which refers to one or more horizontal lines, and a reference elevation will usually be on the top.

The witness stake, usually ½ × 2 in. will have information on the distance from the tack to a part of the structure, such as the center of a pipeline or the corner of a building. It will also have information on the distance above or below the top of the stake to some part of the structure, such as the invert of a pipe or the top of a floor, foundation, or wall. The witness stake will be close to the grade stake to which it refers.

Slope stakes, with ½ × 1½ in. tops, will be set at the point on the ground where the slope of an embankment or excavation meets the original ground surface. A study of Figure 20.29 will help you understand the various stakes.

Of course, all construction stakes must be protected from disturbance until construction is completed. If it is necessary to destroy some stakes during construction, offset stakes may be required. Before moving any stake, confer with your superior.

20.46. Standard Operating Procedure (SOP)

The letters SOP stand for "standard operating procedure," meaning that it is the customary manner of doing a task. When you hear someone say "that's SOP" he means that that is the way we do it here.

20.47. Standby Equipment

Most sewage treatment plants, pumping stations, processing plants, hospitals, and other facilities will have standby equipment, which will start up automatically when there is a general power failure. The equipment will be either gasoline or diesel powered, and most will be set up to start at a given time each week and run for a period long enough to keep the batteries charged. On large projects, the standby equipment usually supplies only enough power to keep the vital equipment running.

Like any other permanent equipment, the standby units will be bolted to permanent bases in the manner shown on the plans.

FIGURE 20.30. A standpipe-type water tank. Such tanks are placed on the highest hill available to produce pressure in the water distribution system.

20.48. Standpipes

Referring to Figure 20.30, this type of water tank is called a standpipe, as it resembles a pipe standing on end. The tank may be open or roofed. The tank will sit on a concrete ring wall under its sides. Inside the ring, the earth will be thoroughly compacted and usually slightly higher at the center than at the edges. Some specifications require a layer of sand under the floor plate. (See Section 24.8.)

The wall and roof, if any, will be made up of steel sheets welded together.

These tanks are usually erected by subcontract, but the concrete ring and earth support inside the ring will be placed by the contractor's own crews. The concrete ring will be formed for pouring and have reinforcing bars. The base of the walls may have an angle for anchor bolts, although many are built without them. If bolts are used, the holes in the angle must be large enough to allow contraction or expansion when the tank is empty.

The top of the concrete ring should be a few inches above the ground to keep moisture from the steel.

20.49. Star Drills

The star drill is round, with the cutting ridges radiating out from the center somewhat like a star. The ridges will be of hardened steel for cutting into rock. The smaller (½−1 in. diameter) drills can be used by hand with a heavy hammer or in an electric hammer. The larger ones (1 in. and up) will be used in a jackhammer or drilling rig. These will have a hole in the center of the rod, through which compressed air or water is passed to cool the drill and carry out the rock particles and dust. With power drills, if the drill point is not kept cool, the steel will lose its temper, and the cutting edges will be dulled.

20.50. Steam Jennies

The little steam jenny is a very useful piece of equipment, particularly in freezing weather. The unit is a small high-pressure steam boiler which produces a small amount of steam at very high pressure. In the summer they are excellent for cleaning equipment, but in the winter they have many uses. The heat from the steam can warm up equipment that is hard to start. The steam under pressure can clear frost, snow, and ice out of forms before concreting starts. They can thaw out frozen pumps and water lines and take the frost out of the subgrade. They can be used to warm on-site aggregate before it is put in the concrete mixer.

The jenny may be on a four-wheeled cart or it may be on skids. It will have a 2−4 ft^3 high-pressure boiler, a kerosene (or other fuel) heating element, up to 100 ft of high-pressure steam hose, and a metal control nozzle.

20.51. Steel Pipe

Steel pipe may be hot-rolled into one piece, or it may be of welded sheets, especially if it is a larger-diameter one. The joints will be welded in the field, and the specifications may require that the welds be x rayed to check the quality.

As steel pipe is subject to corrosion underground, most specifications will require that it be wrapped with moisture-resistant paper or fabric. As the welding will destroy the wrapping for several inches on each side, the pipe must be rewrapped after the weld has been completed. (See Figure 24.12.) Figures 24.9 through 24.12 show a 30-inch steel water pipe being installed. Steel pipe is flexible enough that it can be assembled on the timbers spanning the trench. After several lengths have been welded, the wrapping

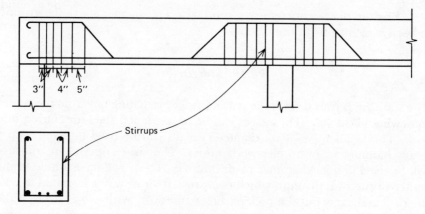

FIGURE 20.31. Stirrup bars.

has been repaired, and the pipe inspected, a crane lifts the pipe to remove the timber, and the pipe is lowered into the trench.

20.52. Steelworkers (See Section 10.13.)

20.53. Stirrup Bars

Referring to Figure 20.31, the stirrup bars are placed where the shear stresses on the beam are greatest. The stirrups will be ¼- or ⅜-in. bars, bent as shown and wired to the upper and lower bars so that they will not move during concreting. The stirrup bars will come in bundles marked for the beam or girder for which they are intended.

20.54. Stone Masons

Good stone masons are hard to find, as the demand is decreasing. Most projects now use precast concrete panels to simulate stone, and besides, there is a long apprenticeship to become a first-class stone mason. However, there are projects that call for natural stone, and such masons will work pretty much on their own, as there will be few people on the site who can tell them how to do their job.

20.55. Stop Orders

As a foreman, you must follow the instructions of your superior as to the manner of doing the work and the quality of the materials used. Sometimes the engineer or his inspector may object to the quality of the material or the method of construction and issue a stop order. If the engineer or inspector questions the quality of your work or materials, confer at once with your superior and be guided by his instructions. If by following those instructions a stop order is issued, let the engineer and your superior work it out—you are only following instructions.

20.56. Stringing Tile

Restrooms and laboratories will have most of their wall surfaces covered with glazed tile. A coating of mortar will be spread over the wall surface, and the tile will be set on it. To keep the width of the joints between the tiles uniform, many masons will place a heavy mason's cord between the tiles as they are being set. When the tiles are set, the cord will be removed, and the joint will be filled with white or colored mortar.

20.57. String-Line Grade Controls

The string-line control developed by the CMI Corporation of Oklahoma City, Oklahoma, is a very useful tool. The system is a string line supported by metal stakes and brackets along one or both sides of the roadway and set at a predetermined grade. A sensor device attached to the roadway equipment has an arm that extends out to the string line and touches it very lightly. Any increase or decrease in the pressure will actuate the controls and correct the grade.

An operations manual comes with the equipment, and the directions must be followed exactly. Once the grade line has been set, it must not be disturbed. The equipment must be serviced regularly to keep it in proper working order.

20.58. Structural Steel

The steel units that make up the frame of a structure are called structural steel. The units come in various shapes, as shown in Figure 20.32. The units used to be held together by rivets, but that practice has given way to the use of high-strength bolts or welding. When bolted, the bolt holes will be lined up by use of driftpins, and when welded, the units will be held in place by steel clamps. Most jobs will be bolted, with the nuts being brought up tight by use of torque wrenches. (See Sections 5.18 and 21.22.)

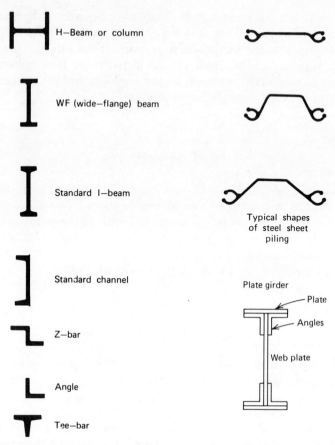

FIGURE 20.32. Standard shapes of structural steel. These shapes are standard for most mills.

20.59. Stucco

As a foreman on a civil engineering project, you would seldom be involved with stucco; however, it is used on occasion for architectural effect on administration buildings. If you are involved, the general procedure is as follows. The stucco is a cement plaster applied to masonry or concrete surfaces. It is similar to mason's mortar, and is mixed and applied in much the same way as plaster. The surface to receive the stucco should be damp, but not wet. The material is applied with mason's trowels, and the surface may be smooth or roughened. The surface may be textured by strokes of the trowel, and sometimes a stipple effect is obtained by tapping the surface with damp burlap. The surface should be kept damp until the stucco has set. The application of a large amount of stucco would be under a masonry foreman.

20.60. Subcontractors

A large project may have several subcontractors, such as plumbers, electricians, sheet-metal workers, steel erectors, painters, and others. They will all have their own foremen. The subcontractor's and the general contractor's workers are supposed to work in harmony, but sometimes they do not. If you have problems with the subcontractor's crews, confer with your superior and be guided by his instructions.

20.61. Subgrades

The foundations of all permanent structures will be placed below the natural grade of the ground after the area has been excavated to the desired depth. The earth at this depth will be called the subgrade. The foundations for structures, earth embankments, roadway base courses and lining, if any, and/or pipe are all laid on subgrade which has been prepared to the proper line and grade and fully compacted.

The most desirable subgrade is bedrock or hardpan, but that is seldom present. Most subgrades will be in earth which is not always of uniform density and must be prepared by rolling or otherwise compacted. The compacting process will uncover soft areas which must be removed, and the void must be filled with acceptable material properly compacted. When compacting by rolling, soft spots can be detected by movement of the earth in front of the roller as it proceeds. With vibrators, the unit will sink further into the soil at soft spots.

It is difficult to develop a stable subgrade through swampy areas. Referring to Figure 20.33, this method was used to build a new road in an industrial area, and it never settled, although it has been in use for many years. There was a gravel pit which furnished the area with crushed gravel

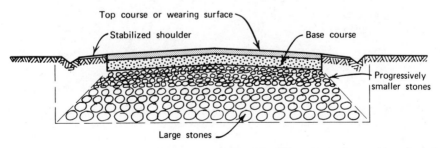

FIGURE 20.33. Stabilizing a swamp area for roadway. Large stones are dropped into the soft earth and allowed to sink. As the area stabilizes, smaller stones are used until the elevation of the bottom of the base course is reached. The subgrade is then rolled, and additional stones are added until the area is stable.

that had a large pile of stones too large for the jaws of the crusher. These large stones were trucked to the site and dumped into the soft swampy area. The first few loads sunk out of sight, so slightly smaller stones were added until a base was built up. When the base became stable, graded stones were deposited up to the elevation of the bottom of the roadway base course.

Note that the first few layers of large stones were carried out far enough to give a 45 deg slope, which reduced the pressure per square foot on the soil.

Years ago, it was customary to lay a thick layer of tree branches and brush over swampy areas to act as a filter to keep the sand and silt from filtering up into the roadway base. Today, better results are obtained by placing a blanket of canvas, plastic, or other fabric over the subgrade and placing the road base courses directly on it.

If a subgrade is prepared and then left exposed to the weather for some time, the top 1 or 2 in. will become unstable. If it is too wet, the mud can be bladed off and that much concrete added to the depth of the foundation or slab. If time is no problem, the area can be left to dry out and then be rerolled. Lime can be spread over the area to hasten the drying. If the subgrade is too dry, a fine spray of water can be applied and the area can be rerolled.

FIGURE 20.34. Preparation of area for a sewage treatment plant. The subgrade had hard shale in one end of the excavation and soft clay in the other. The smooth area in the background of the photograph is the top of the leveled-off shale. At the right-hand side and in the foreground, the clay has been removed and the area backfilled with gravel. The gravel was placed in 8 in. layers and compacted. The depth was about 7 ft and was deemed to be less expensive than using piling and grade beams.

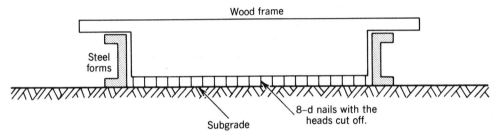

FIGURE 20.35. When preparing the subgrade for a concrete highway slab, this template is a useful tool. The steel forms are set to the required line and grade, and the template is slid along them as shown. If the nails touch the grade and hold the template above the form, the grade is too high. If the nails do not mark the grade, it is too low. If the nails just make a slight mark on the grade, it is at the correct elevation.

One of the most difficult problems in preparing a subgrade occurs when the soil in the area has quick changes in classification. Part of the area can be shale with a high load-bearing capacity, and then it can suddenly change to heavy clay, which has a low load-bearing capacity. These quick changes are not always detected by the test borings. The decision on what to do will be up to the engineer. The clay may be removed and replaced by run-of-bank gravel, but that is usually expensive. Piling may be required, or a large-diameter soil auger may be used to bore holes to stable soil, and the hole will be filled with concrete.

Hard clay will sustain a reasonably heavy load for a short time, but will eventually yield and cause settlement of the load. If the clay is to be removed, the most economical way is by pan (scraper), if there is room for maneuvering and a place to deposit the load. If not, the front-end loader or backhoe is used, and the clay is trucked to a point of deposit (see Sections 4.32, 19.17, and 23.6). (See Figures 20.34 and 20.35.)

20.62. Substantial Completion

Civil engineering project contracts will have a clause requiring completion of the work by a specified date, and if the work is not completed by that date, the contractor will be assessed a stated amount of money for each day of delay. Under such conditions, the contractor may claim substantial completion. Some courts have said that a project does not have to be absolutely complete to meet the terms of the contract. For example, a bridge may be complete to where it can carry traffic, but still require painting, riprap, and landscaping and cleanup. In each case, the project would be fully usable by the owner or the public.

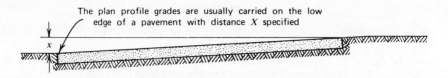

The plan profile grades are usually carried on the low
edge of a pavement with distance *X* specified

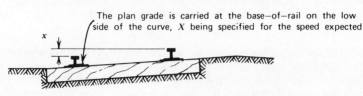

The plan grade is carried at the base—of—rail on the low
side of the curve, *X* being specified for the speed expected

FIGURE 20.36. Superelevation. The plan profile grades for highways (top) are usually carried on the low edge of a pavement with the distance *X* specified. The plan grade for railroads (bottom) is carried at the base-of-rail on the low side of the curve, *X* being specified for the speed expected.

20.63. Substitution

For a discussion on the substitution of materials or equipment, see Section 16.9.

20.64. Superelevation

Figure 20.36 will show you what superelevation is. Without this higher elevation on the outside of a curve, the vehicles would slide off the pavement or a railroad car would topple over, particularly at high speed.

The change as a vehicle approaches a curve must be smooth and gradual; otherwise, the vehicle will sway and the driver will lose control. On a passenger train, a person could be thrown to the floor.

20.65. Superintendents

The superintendent is the overall boss of the contractor's organization and is directly responsible to the owner for the progress of the work. On large projects, he may have one or more assistant superintendents or general foremen. You will be advised as to who will be your immediate superior.

20.66. Supplies

Availability of the required supplies is very necessary to keep a construction crew in full production.

All tools needed for the day's work should be on hand and in good order. All equipment should be lubricated, full of fuel, and in good working order. Material required should be near at hand, or arrangements should be made to have it delivered as needed. When material is being delivered continuously, as is earth for embankments, Portland cement, bituminous concrete, or any such material, there should be at least one truck waiting to unload. The lack of any of these can affect your crew's production, so if you are having such problems, you should confer with your superiors in an attempt to improve delivery.

FIGURE 20.37. The Swiss hammer shown is used to determine the strength of concrete. The hammer has a plunger with a very strong spring. The plunger is set in the cocked position, and the lower projection is placed against the concrete. As the unit is pressed down, the cocking mechanism releases the plunger. The degree of resistance by the concrete is a measure of its strength. Several readings are taken, and the average is referred to a chart that gives the strength in pounds per square inch. The tests vary about 20% from what a test cylinder would show for the same concrete. Engineers use the Swiss hammer to determine when to allow the contractor to remove support forms.

20.67. Surge Tanks

The large conduits that carry water from a reservoir or lake to the turbines of a power plant contain a large volume of water moving at considerable speed. If a turbine is damaged and requires an immediate shutdown, the valve to the surge tank is opened as the value to the turbine is being closed. This allows the water to surge up into the tank and slow its velocity. The tank will be similar to a standpipe, with the size depending on the volume of water. Without the surge tank, the tremendous pressure could rupture the conduit.

20.68. Swiss Hammers

The swiss hammer is used to check the hardness, and thus the strength, of recently poured concrete. The device has a metal cylinder and a short steel plunger. The plunger is forced back against a strong spring and locked in place. As the device is pushed against a concrete surface, the locking is released, and the plunger is forced against the concrete. The rebound is measured on a scale on the cylinder, and the amount will indicate the approximate strength of the concrete. The information is used to determine if the concrete is strong enough to allow the removal of the forms holding a beam or if a slab is strong enough to allow traffic. (See Figure 20.37.)

20.69. Switch Points

Referring to Figure 7.16, the switch point is the short tapered section of track that moves as the switch is thrown. It allows a smooth transfer from one track to the other.

21

21.1. Tamping (See Section 4.32)

21.2. Tanks

On civil engineering projects, tanks can be built of various materials and be of various shape and sizes, depending on their expected use. They are used at sewage treatment plants, in water supply systems, for storage of fuel and other liquids at processing plants, and for storage of gases.

Steel Tanks

Steel tanks are used both above and below ground. Those underground, in areas of high groundwater, must be anchored by bolts or steel straps to rock or concrete base heavy enough to resist floating when the tank is empty. The tanks may be protected by a bituminous coating and/or a cathodic system. Small steel tanks will come complete, while large ones will be fabricated on the job by welding. On very large tanks, the welding will be done by a machine. Most states require that the welders by certified as competent.

Large tanks above ground and bolted to a foundation will have provisions for expansion and contraction. The plans will give the details. Tanks in which the length is greater than the diameter will, when standing on one end, be referred to as "standpipe.' Those in the air on legs will be called "elevated tanks." Such tanks are usually in a water supply system to maintain pressure on the line. The steel legs will have guy rods to stabilize the structure, and each rod will have a turnbuckle to adjust the tension.

Steel tanks will be painted with a prime coat and one or more other coats. Most specifications will require that each coat be of a slightly different shade of color so that each coat can be checked when being applied. A total thickness of paint will be required, and the inspector will have a gauge to check the final application.

Steel tanks for holding gas will usually be within an open-top concrete tank containing water. The steel tank will have no bottom. The edges will extend down into the water, providing a seal to contain the gas. The steel tank floating in the water will provide pressure on the gas main. There are other types of gas tanks, but this is the one I am most familiar with.

Steel tanks will be erected by ironworkers or steel erectors, and they will have their own foremen.

Concrete Tanks

Concrete tanks may be round or rectangular. They may or may not have provisions for expansion and contraction. The size and position of the reinforcing steel will be shown on the plans. Some round tanks will have no reinforcement in the wall, but will have high-strength wire wound around the outside and covered with Gunite. The wiring will start at the bottom of the wall where the beginning end will be attached to an anchor. A special machine will wind the wire at the required spacing and tension with one continuous strand, and when complete, the end at the top will also be anchored to maintain the tension.

When the sides or the circumference of a concrete tank are over 40 ft, it will be necessary to splice the reinforcing bars. The splice should be what is specified, or about 30 diameters of the bar. The splices should be staggered so as not to be one above the other. Many 12 in. thick walls will be 8–10 ft or more in height, and if so, the concrete should be poured through a tremie and be well vibrated around the bars.

Wooden Tanks

It used to be that most elevated tanks were made of tongue-and-groove planks bound by steel rods, but today they are mostly steel. However, occasionally a wooden tank is used for architectural reasons. If you should be involved with one, be sure the tongue-and-groove joints are up tight and that the joint between the wall and floor is exactly as shown on the plans.

All elevated tanks must have the legs well secured by guy rods, and the anchor bolts must be well set into the footing to resist tension when high winds put pressure on one side.

21.3. Tapping Sleeves and Valves (TS&V)

Tapping sleeves and valves are used when a new lateral pipeline is to be attached to an existing water main that is under pressure. Figure 21.1 shows

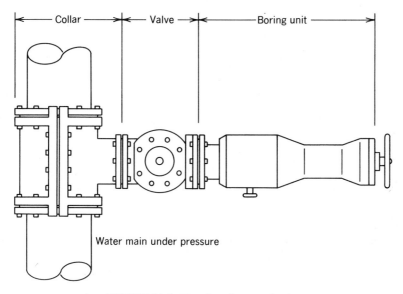

FIGURE 21.1. Tapping sleeve and valve.

how the unit works. The TS&V is attached to the main as shown. The exterior of the main must be clean and smooth before the unit is installed. After the unit has been installed and all the bolts are tight, the new valve is opened wide, and a drilling device inserted through the valve to the wall of the main where it will drill a hole through the wall and retain the piece of pipe wall cut out. The device is then backed out beyond the new valve. The value will then be shut, and the device will be removed, leaving the new valve in place with water pressure against the side next to the main. The lateral pipe can then be installed, and when completed, the water is turned on.

21.4. Taxiways (Airport)

Taxiways, like runways, are constructed similarly to highways, except for their depth. The various base and top courses are constructed in layers, the number of which depends on the traffic. The plans will show the composition of each course and its depth. (See Section 19.23.)

21.5. T-Beams

When a reinforced concrete floor system is poured in one piece, that is, the beam and slab are poured at the same time, the design is called "T-

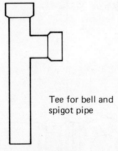

Tee for bell and
spigot pipe

FIGURE 21.2. Tee.

beam" construction. Referring to Figure 17.1, the beam is poured up to the bottom of the slab, and the concrete is allowed to settle. After a few minutes, the pouring resumes to complete the slab. Because of the greater depth of concrete at the beam, there will be more settlement, and the wait will prevent cracking at the bottom of the slab. The pouring of the slab should resume before the concrete in the beam starts to set, in order to ensure a proper bond, as the full depth of the beam is considered to be from the bottom of the beam to the top of the slab. Good vibration hastens the settling of the concrete. If gain water should appear in the form as the beam concrete is settling, it must be removed before the slab is poured. (See Section 8.2.)

21.6. Tees

Pipes such as that shown in Figure 21.2 are called "tees." (See Section 24.31.)

21.7. Terrazzo

Terrazzo is a concrete in which the large aggregate is marble chips. It is used as a floor surfacing in commercial and industrial buildings and sometimes in residential work, particularly in the South.

The mix is spread over a concrete base to a depth of 1–2 in. and troweled smooth. The mortar may be colored. The mix is allowed to set, and then the surface is ground with a carborundum stone to a glasslike finish. The marble chips will be exposed to make up about 85% of the surface. After grinding, the surface will be cleaned and polished. Most such installations will be divided into 3–4 ft squares by use of metal strips. The strips improve the appearance and reduce the chance of cracks in the surface.

Once the surface has been ground, it should be cleaned, polished, and sealed to prevent grime from getting in.

The terrazzo should be cured slowly, the same as any mortar or concrete. All work in the room, such as installing partitions and painting, should be completed before the surface is ground. The surface should be protected by a layer of sand or heavy building paper until the building is ready for use.

21.8. Test Cylinders

The engineer or his inspector may take test cylinders of the concrete, or laboratory personnel may do the job. In either case, you should cooperate with them to get the test done quickly and avoid delaying the concrete pouring.

21.9. Test Holes

The testing of the soil is usually completed before the construction starts. However, some may be required during the work, and you may be assigned to assist by boring with an earth auger or digging a hole by hand or machine. What is needed will depend on the type of test to be made. The actual testing will be done by the engineer or his assistants. (See Figure 21.3.)

21.10. Testing

The testing of materials and permanent equipment is the responsibility of the engineer and his inspector. Your only concern is to see that you do not use any material requiring testing until it has been approved. If you are

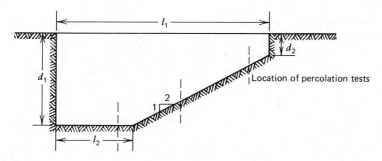

FIGURE 21.3. Test hole dug by backhoe. The width will depend on the width of the backhoe bucket.

FIGURE 21.4. Most specifications will require that a certain number of pipe sections be tested to destruction. The pipe is placed in the testing machine as shown, and pressure is applied by the hydraulic jacks on top. Pressure is applied until the pipe cracks, at which time the total pressure is noted. From that, the pressure per linear foot is derived. The pounds per linear foot required to crack the pipe must equal or exceed the specified amount.

working on a subgrade or embankment that requires testing, try to have the personnel conducting the tests on hand when needed so you can avoid delay of the work. (See Figures 21.4 and 21.5.)

21.11.   Test Loads

Test loads are usually conducted by the engineer or architect before the design is completed. However, the contractor is sometimes required to make such tests, so you should be familiar with the procedure.

Test loads on the soil are made in the manner shown in Figure 21.6. The timber mat will be from 3 to 4 ft. square and the amount of test weight will depend on the pounds per square foot (psf) required. The steel wire must be taut and measurements should start with the loading and reported every 100 pounds or so until the full load has been placed. Usually the full load is left on for at least 24 h and then unloaded in the same increments as the loading with a reading of the gauge at every increment. Mounds of earth should be placed to keep surface water from entering the excavation and any seepage into the pit should be removed at once to keep from softening the earth under the mat. In rainy weather a tent or other cover should keep the rain out of the pit.

Test loading of a pile is done as shown in Figure 21.7. Here again

FIGURE 21.5. A method of testing a cast-iron pipe for a water main. Both ends are sealed, and water under pressure is forced into the pipe through one of the end plugs. The pressure is usually built up to twice that expected of the pipe under normal use. The specifications will indicate the pressure and the time it must be held to be acceptable.

the test wire must be kept taut and readings must be taken at stated intervals during both loading and unloading.

21.12. Test Piles

Where there is to be considerable piling on the job one or more test piles may be required. The tests are conducted in various manners with one method shown by Figure 21.7. The method to be used will be indicated by the engineer. If your crew is assigned to assist just follow the engineer's instructions. (See Figures 21.8 and 21.9.)

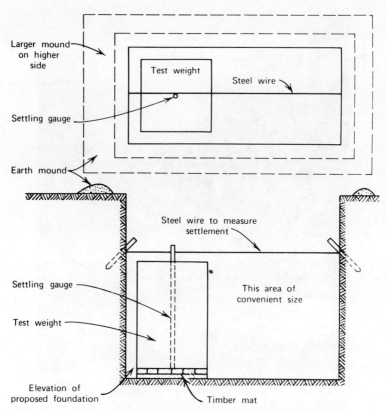

FIGURE 21.6. Testing bearing capacity of soil. A method of taking soil-bearing value tests.

21.13. Thermometers

Thermometers are used on the job for several purposes, such as checking concrete, asphalt mixes, the weather, and the water temperature for mixing concrete or mortar in cold weather. Most such checking will be done by the engineer or his inspector, and will have the various types of thermometers used.

21.14. Thrust Blocks (See Section 3.20.)

21.15. Tiebacks

Figure 21.10 shows what a tieback is and how it works. If constructed in soil, the "deadman" must bear against solid earth of sufficient strength to

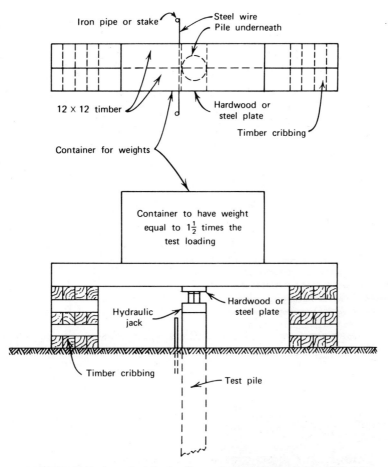

FIGURE 21.7. A method for loading a test pile using a hydraulic jack.

take the pressure, and the backfill around it should be well tamped. If the installation is to be permanent, the tie rod may be covered with a protective coating. If so, it must be protected from damage. If rock bolts are used, the void around the rod may require filling with mortar. (See Section 19.16.)

There should be a large cast-iron or heavy cutwasher between the waler and the nut at the end of the rod.

21.16. Tie Plates

Tie plates are heavy cast-iron plates set on railroad ties to carry the rails. The plates spread the load and prevent the base of the rail from cutting

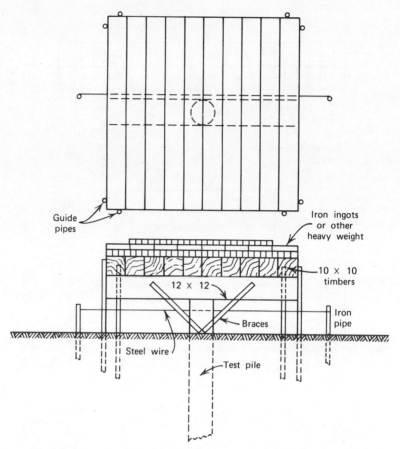

Guide pipes

Iron ingots or other heavy weight

10 × 10 timbers

12 × 12

Braces

Iron pipe

Steel wire

Test pile

FIGURE 21.8. A method for using a test pile with a loaded platform.

into the tie. The plates will be set on the ties loose, and the rails will be set on them before the plates are spiked to the tie. The rail must sit in the grooves in the plate before the plate is secured to the tie by special railroad spikes. A special hammer with a long, narrow head is used to drive the spikes.

21.17. Tile

Tile work will be done by a subcontractor or, if done by the contractor's own forces, under a mason foreman.

There are many different kinds of tile and many different uses. The tile may be of baked clay, cement, stone, rubber or other composition, or

FIGURE 21.9. A 200-ton hydraulic jack under a loaded platform testing steel H-beam pile.

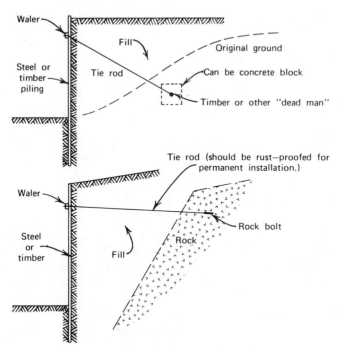

Waler

Fill

Original ground

Steel or timber piling

Tie rod

Can be concrete block

Timber or other "dead man"

Tie rod (should be rust—proofed for permanent installation.)

Waler

Rock bolt

Steel or timber

Rock

Fill

FIGURE 21.10. A method of placing a tieback in earth by use of a "deadman" and in rock by use of a rock bolt.

plastic. The baked-clay units are used as roof tile, and the cement is used as wall tile, in the form of rectangular blocks, or as decorative walk and patio tile, usually 2 in. thick and 12–18 in. square. The rubber or composition tile is used on interior walls of kitchens, restrooms, and other special areas, and sometimes on the floors of these areas. The plastic tile is used in similar areas as the rubber or composition tile.

The proper installation of tile requires a considerable amount of skill. If you expect to supervise such work, it is recommended that you work at it for a few years.

21.18. Timber Construction

Timber construction today is mostly confined to the interiors of buildings, for partitions, floor and ceiling joists, roof boards, and stairways. The roof structures of auditoriums, gymnasiums, and churches may have timber trusses constructed of laminated beams or arches. Such heavy units will have cast-iron or steel-plate end-bearing devices. The units will come complete, ready for installation, and most will be protected by a covering of heavy building paper, which should not be removed until the unit is set in place. Some bearing devices will be plates, while others will be large-diameter bolts. Care must be taken to see that the bearing devices are set at the correct elevation and level so that the beam or truss will be level at both ends and vertical. The bearing will usually be detailed on the plans. The partitions and joists will be set in the same way as for any frame construction. Laminated beams or arches that are prefinished must be handled and stored so as not to damage them because refinishing is costly and they are hard to match. When hoisting, use rope or canvas slings, not chain or cable.

21.19. Timekeeping

Timekeeping is usually the foreman's responsibility, although on very large projects an office man may be assigned that duty. Being paid the right amount and on time is very important to the workers.

Usually, the contractor will furnish a time book that conforms to his method of timekeeping. The book will have columns for the worker's name, classification, pay rate, and hours of both regular and overtime work. You must see that all these items are correctly recorded, as any problems with his pay upsets the average worker. You will have much better morale in your crew if they know you are concerned that they get paid the correct amount and on time.

21.20. Tools

All tools must be of the proper kind and in the proper condition if your crew is to work at top efficiency. The tools must be clean, sharp, and well lubricated as required. Each worker must know how to use his tools properly and safely. A worker using power tools must keep alert and have the strength to control the unit. All tools should be cleaned at the end of the day's work and properly stored until needed again.

21.21. Topsoil

Most projects will require that the topsoil be removed from the area where construction is to be undertaken and stored at a designated area for use in landscaping after construction. When being removed, care must be taken to pick up only the topsoil and not the underlying earth.

After construction, the topsoil can be easily distributed by use of a dump truck with a restricted tailgate. With a little practice, the tailgate can be opened just enough to allow the spreading of the 2–4 in. usually required. When large amounts of soil are involved, scrapers may be used.

21.22. Torque Wrenches (Torsion)

Torque wrenches are used to tighten the bolts on structural-steel connections. The torque is measured on a gauge on the handle. As the handle bends, the pressure is indicated on the gauge. The idea is to get the nut as tight as possible without stripping the threads. The specifications will indicate the maximum pressure allowed.

21.23. Training

Many contractors conduct training classes for their foremen and superintendents. Some classes are only short lectures on generalities, while others will have extensive classroom instruction and hands-on experience. Some even have written tests. It is recommended that you take as many of these courses as possible to learn your employer's methods. A thorough study of this book will help you to a better understanding of his methods. Not all contractors do each item of work in the same manner, although their methods will be similar. Your employer will expect you to do the work in his manner.

Some community colleges conduct construction classes in both the daytime and the evening, and the classes are highly recommended. If you can attend a technical school or college, that is even better.

21.24. Trash Racks

Trash racks prevent sticks, tough fiber items, and other debris from getting to and damaging the pumps in sewage treatment plants and powerhouses. The rack may be of timber or steel, will be set at about a 30 deg angle from horizontal, and will have hand- or power-operated machinery for removing the debris. The rack will usually come in one or two pieces, ready to install the same as any other machinery. The plans will give the details. The rack must be well bolted to the structure so it will not move when the trash is being removed.

21.25. Trees

Construction can be rough on trees and shrubs, and when they are to remain part of the landscaping, their damage or destruction can be costly for the contractor. The specifications or plans will indicate which trees and shrubs are to be saved, and some method of protection should be installed before the construction starts. One way to protect them is to erect a temporary fence or line of posts. Trees can be protected by placing 2 × 4s vertically against the trunk and binding them with wire. A mound of dirt around a tree will keep moving equipment away.

Many trees have been destroyed by trenching too close and cutting out too much of their root systems. Usually, if over a third of the root structure is destroyed, the tree will be damaged, if not killed. If a damaged tree lives through the next season, it will survive.

It takes a long time to grow a tree, so it is good to save them if at all possible.

21.26. Tree Damage (See Section 21.25.)

21.27. Tremies

By use of a tremie, it is possible to pour concrete underwater. Figure 21.11 indicates how a tremie works. Once the pour starts, the bottom of the pipe must be kept well within the mass of concrete; otherwise, the concrete will rush out and the action will wash the cement out. If the pipe is inadvertently lifted out of the mass, it must be removed, a new plug must be inserted, the end must be set back in the mass, and then the concrete pouring can be started again. The concrete should have a plastic consistency and be allowed to pile up several feet before moving the tremie. The sidewise movement of the tremie should be slow, and concrete should be poured as it moves in order to keep the end well within the mass. A depth gauge will

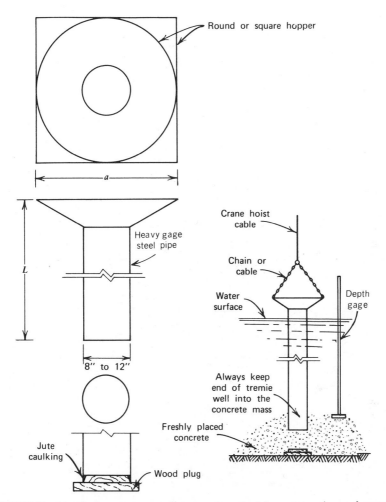

FIGURE 21.11. Tremie for concrete underwater. A method for constructing and operating.

show how the mass is spreading and when the desired depth has been reached. The finished top will be uneven, and an inch or more on top will have the cement washed out and will have to be removed. If the concrete is poured within a cofferdam, it should be poured for the full width, out to the sheeting. If 2 or 3 ft is poured in a cofferdam, the remaining height can usually be poured in the dry after pumping out the water. Often, the footing of a bridge pier will be poured by tremie, then the cofferdam will be pumped out, and the pier forms will be placed in the dry. In addition to the roughness on the top of the tremied concrete, a key should be cut to a depth of several inches to augment the resistance to sliding. If tremie use is anticipated, the plans will indicate how the top is to be treated. If

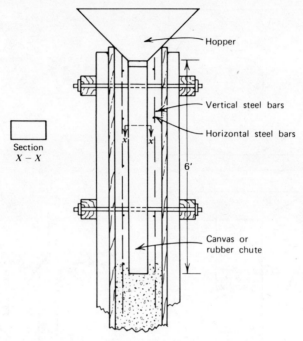

Section
X – X

FIGURE 21.12. Tremie for concrete placed in forms. Ready to receive the concrete.

the tremie is used because the cofferdam cannot be dewatered, it is recommended that a diver place heavy dowels to increase the bond.* The concrete should set at least seven days before dewatering is attempted. (See Figure 21.12.)

21.28. Trenching

Most trenching on civil engineering projects is done by backhoe (pullshovel), although trenching machines and power shovels are used. (See Section 6.9.)

Trenches are seldom less than 24 in. wide, as that is about as narrow a space as a worker can function in efficiently. The width may be 5 or 6 ft or more, depending on the job.

The greatest danger is from the collapse of a trench wall. There are strict laws governing trenching in most states, and OSHA also gets involved. The deeper the trench, the greater the danger. A collapsing trench can injure or kill a worker and can push a pipe out of line and even damage it.

* Use ¾ in.$^\phi$ bars, 36 in. long for dowels.

Observe the walls of the trench as work starts to see if they remain stable or are inclined to slump. One of the first signs that a trench wall is about to collapse is a line of cracks on the surface parallel to the trench. Also, a bulge in the wall indicates that moisture is lubricating the soil and making it unstable. When the excavated earth is placed too close to the trench, the weight can cause the wall to collapse. The near edge of a pile of excavated earth should be at least a distance away from the trench equal to the depth of the trench at that point.

There are several methods of preventing a wall collapse. (See Section 20.16.) If large stones or boulders are exposed in the wall of the trench, they should be removed or braced to prevent them from falling where they could injure a worker or damage the pipe. (See Section 20.19.)

Water seeping into the trench from the wall or flowing in from the surface can soften the subgrade. If there is surface water, it should be directed away from the trench by ditching or building a ridge of earth. If there is seepage from the walls, a small ditch along the base of the wall will carry the water to a sump hole where it can be pumped out. Heavy seepage from the wall will require sheeting to keep the wall from collapsing.

If you should be so unfortunate as to have a worker buried by a collapsing trench wall, action must be taken *immediately* to uncover his head and chest so he can breathe while being dug out.

Someone should be assigned to observe the walls for further collapse and should warn the rescuers.

When trenching in earth, small pockets of quicksand are sometimes encountered. Such pockets can be 4 or 5 ft to 20 ft or more in size and are exposed as the trenching cuts across the area. Removal of the quicksand could cause the trench walls to collapse. If the pocket is above the subgrade elevation, a decision must be made about either letting the wall collapse and excavating the material or placing sheeting to hold the earth and let the water drain out. If the pocket is at or below the invert level, the following procedure will stabilize the area: No. 2 and No. 3 crushed stone dumped into the quicksand will sink out of sight, but continuing to place the stone will force the quicksand up into the trench where it can be removed in order to stabilize the area.

Another condition encountered when trenching involves areas of water-bearing clay and silt. Such areas are difficult to dewater by well points as the groundwater is suspended between the clay and silt particles and flows very sluggishly. The subgrade will be like jelly or stiff mud. Under such conditions, the following procedure has been found to work. Using a ¾ yd backhoe, the earth is excavated to the depth of the bucket by lowering the bucket into the soft earth and pulling it forward slowly. As a void opens behind the bucket, well graded run-of-bank gravel is poured in and rammed down tight. As the bucket continues to move forward, the gravel is poured in behind it to keep the void full. (See Figure 21.13.) The result is at least

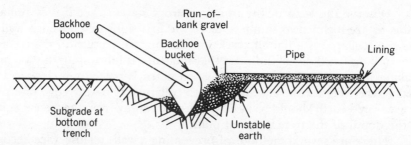

FIGURE 21.13. As the bucket is pulled forward in the unstable earth, run-of-bank gravel is dumped in behind and compacted. The depth of gravel will depend on the soil condition.

When proceeding with normal trenching and pipe laying and an unstable area is uncovered, proceed with the method shown above until stable ground is reached. It may be 10 or 100 ft.

2 ft of well compacted gravel lining. The sewer pipe is installed as quickly as possible, and gravel backfill is placed around and over the pipe to a depth of at least 2 ft. The remainder of the backfill is placed to fill the trench and restore the normal pressures in the area.

Both of the above procedures could be termed unorthodox, but I have returned to areas after a year and lamped the sewer, and found them to be on line and grade. After several years, they were still operating satisfactorily.

The following is a checklist for foremen on trenching for sewers:

1. Batter boards must be placed at least 200 ft ahead of the trenching.

2. The cord or wire between batter boards must be kept taut, without sag.

3. The backhoe or trenching machine must deposit the excavated earth at least 10 ft back from the trench.

4. If large stones roll out of the spoil pile, they must be placed at the toe of the pile, away from the trench.

5. As the excavation nears the final depth, the bucket must be kept nearly parallel with the subgrade so that the teeth of the bucket will not cut deep grooves in the earth. Such grooves will prevent even compacting of the subgrade or lining.

6. Keep constant watch for cracks on the surface parallel with the trench, as such cracks indicate that the trench wall is settling and may collapse if sheeting is not installed. If the cracks are small and the surface settlement is slight, the trench walls may be sloped about 20 deg from vertical, which will usually stabilize the bank. Spot sheeting is another way of holding a questionable bank. (See Figure 21.14.)

7. The grade pole should have two markings on it: one for the pipe invert and the other for the top of the subgrade or lining. (See Figure 2.4.)

8. If the soil is okay for subgrading, compact it and bring it to the required grade, frequently checking with the grade pole. If the soil is too

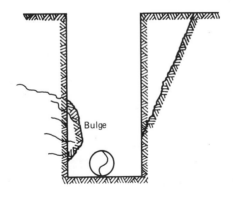

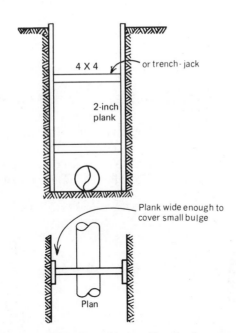

FIGURE 21.14. Trenching problems and treatment.

soft, 3 or 4 in. or more of lining may be required. The engineer or inspector will usually instruct you on the depth of lining they will approve. Of course, additional earth must be removed to accommodate the lining.

9. See Section 17.17 for methods of installation of the various types of pipe.

10. After the section of pipe has been approved by the engineer or his inspector, the trench can be backfilled. A 2-ft cover should have been

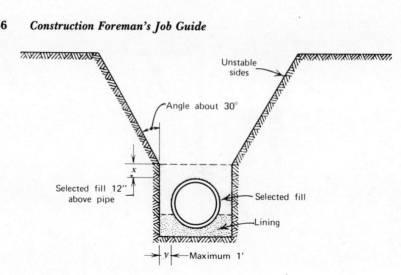

FIGURE 21.15. Trenching in areas of unstable trench walls. A method used when sheeting and bracing are not necessary.

placed over the pipe as soon as it was installed, so the remainder of the fill can be bulldozed in. The amount of compacting required will depend on the location of the line. The specifications will indicate the amount. In open areas, little or no compacting may be required, but a high mound will be placed over the trench to allow for settlement after rains. Trenches under traveled ways will usually require full compaction for the full depth.

11. Not all of the above conditions will be encountered at any one time. However, memorizing the list will help you recognize a possible problem and give you one way of solving it. (See Figures 21.15–21.19.)

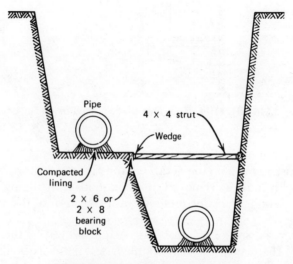

FIGURE 21.16. A two-pipe trench.

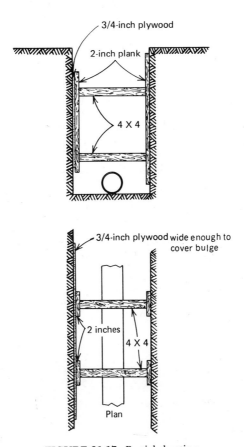

3/4-inch plywood

2-inch plank

4 X 4

3/4-inch plywood wide enough to cover bulge

2 inches

4 X 4

Plan

FIGURE 21.17. Partial sheeting.

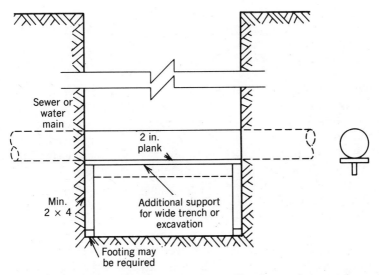

Sewer or water main

2 in. plank

Min. 2 × 4

Additional support for wide trench or excavation

Footing may be required

FIGURE 21.18. Figure shows a method of supporting a utility pipe crossing a trench excavation. If left unsupported backfilling operations might break it.

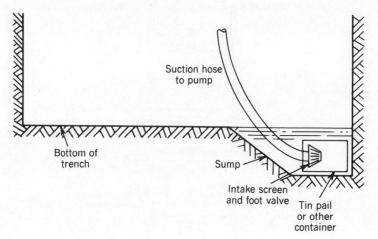

FIGURE 21.19. When the suction pump lowers the water level to the intake screen, the pump will suck air as well as water. This agitation will stir up mud, which may clog the intake. The pail will prevent the agitation.

21.29. Trench Load

Trench load means the weight per linear foot on the pipe caused by the backfilling. The loading depends on the size of the pipe, the width of the trench, and the depth of the backfill. The design engineer will specify the grade of pipe for the different trench conditions. The grade will be stated on the plans, and you must be sure you use the correct one by checking the markings on the pipe sections.

21.30. Troweling (Concrete)

Troweling of concrete is done with hand or power trowels, depending on the area to be finished. In either case, care must be taken not to overtrowel, as overtroweling will break down the mortar and cause dust and hairline cracks when the surface sets. When a concrete slab is poured, there will be moisture on the surface, which will shine. As the concrete starts to set, the surface moisture will evaporate, and the shine will disappear. That is the time the troweling should start. There should be just enough troweling to get the surface in the condition desired, no more.

The hand trowel will be a steel plate with a handle. The plate will be about 4 × 12 in. or larger. The power trowel will have four to eight steel plates, contained in a circle with a power unit at the center. There will be a handle similar to that on a lawn mower, with controls to operate the power unit. The blades will rotate, with the front side angled up slightly. The unit is guided over the surface by the handle, and again, care must be taken not to overdo it.

21.31. Trusses

Trusses can be of steel, timber, or a combination of the two. They are used to carry loads over spans for which beams or girders would be too heavy.

They are used in structures where a large area is required without columns, such as in a hotel lobby, a sports arena, a convention center, or another such enclosed area.

Timber trusses will be held together by plywood or steel gusset plates and steel bolts. The steel ones will have steel gusset plates, which can be bolted or welded. They will be held together by purlins attached to the top member and may have cross bracing between the units. On most jobs, the trusses will be erected by steel erectors, and they will have their own foremen.

Long-span trusses will usually have one end fixed and the other movable on rollers or rockers. The plans will give the details. Timber trusses may be laminated-wood planks, and quite often will come with a varnish finish with a protective paper cover. Such paper should not be removed until the job is complete. Most trusses on civil engineering projects will be heavy enough to require a crane for erecting. The very lightweight wood or metal trusses (see Section 3.4) can be installed by hand, and if so, the work is usually done by the carpenters.

If you are involved in truss erection, study the plans for the method of bearing and bracing. See that the trusses are vertical, and if bolts are used, they must be secured as tightly as possible without stripping the threads.

21.32. Tunnel Liner Plates

Tunnel liner plates are curved steel plate sections which are bolted together to make a circular liner for a tunnel. (See Figure 21.20.) Note that the sections are bolted around the circumference and lengthwise. The tunnel excavation is carried back just short of the width needed to install the plates. The plates are then forced or hammered back into position for installation. This keeps the earth pressure against the plate and prevents any settling of the earth outside. Sometimes the excavation is carried beyond the line for the plate, and a void is left behind the plate after installation. These voids are later filled by injecting grout under pressure through holes in the plates. After the grout is placed, a plug is screwed into the hole to seal it. Voids behind the plates can be located by tapping on the plates with a hammer. If the soil is tight against the plate, there will be a dull thud as

FIGURE 21.20. The heading in a 54-in. diameter tunnel. Note that liner plates are kept close to the excavation. When the photo was taken the workmen had just finished loosening the earth and pulled it back into the tunnel to be loaded on a small two-wheeled cart for hauling out. The two workmen take turns loosening the earth at the heading and loading into a cart. The cart is pushed in and out by another workman. Note the ribs on the liner plate to give added strength. Cement grout will be forced behind the plates to fill any voids.

the hammer hits the steel. If a void exists, there will be a hollow sound. Retapping after grouting will tell if the void was completely filled.

On very large diameter tunnels, the liner plates will be of reinforced concrete, requiring a special machine to install them. (See Section 21.33.)

21.33. Tunnels

Tunneling is dangerous work. If you are to become involved, you should spend considerable time as a worker in a tunnel or get very detailed instructions from your superior. You have to be concerned with ventilation, gas, falling rock, blow-in of unstable earth, flooding from underground springs or broken water mains, and collapse of improper cribbing. A tunnel may be as small as 4 ft or as large as 25 ft or more in diameter. Most are round, but they may be rectangular or almost any shape desired. The small ones will use tunnel liner plates unless they are in rock or very stable earth. The larger ones will use the liner plates, or if the earth is stable enough, a wire mesh may be placed and covered with Gunite. (See Sections 8.21 and 17.29.)

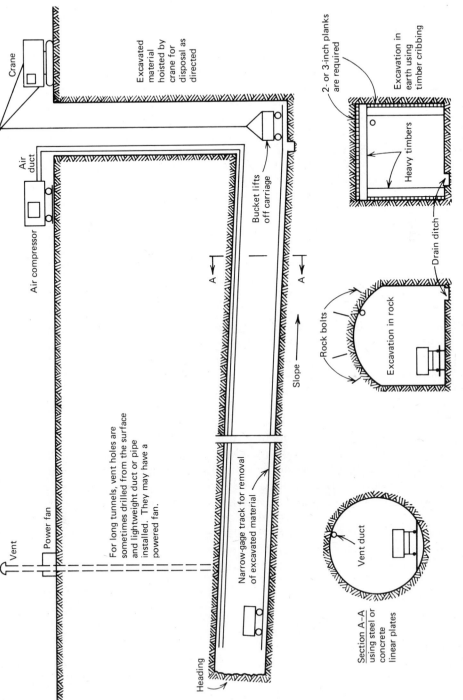

FIGURE 21.21. Tunneling operations vary depending on the material encountered and the contractor's methods. This figure shows the general procedure.

Crane

Excavated material hoisted by crane for disposal as directed

Air duct

Air compressor

2- or 3-inch planks are required

Heavy timbers

Excavation in earth using timber cribbing

Bucket lifts off carriage

A

A

Slope

Power fan

Vent

For long tunnels, vent holes are sometimes drilled from the surface and lightweight duct or pipe installed. They may have a powered fan.

Rock bolts

Excavation in rock

Drain ditch

Narrow-gage track for removal of excavated material

Heading

Vent duct

Section A–A using steel or concrete linear plates

371

372   *Construction Foreman's Job Guide*

If possible, all tunneling should be started at the low end to allow for drainage.

If blasting is required, the tunnel should be well ventilated before workers reenter, as the fumes can cause severe headaches.

When working in unstable earth, material should be handy with which to build a strong bulkhead if necessary. (See Section 3.35.)

The tunnel should be large enough to allow the equipment carrying the excavated material to safely pass a worker standing on the side. If not, there should be safety alcoves spaced along the length of the tunnel. The tunnel should be well lighted so that a worker can see an approaching vehicle.

When excavating at the heading, the liner plates should be installed as soon as possible. Sometimes the workers get careless and get several feet ahead before installing the plates. This practice can result in a collapse of the earth due to an area of unstable conditions or vibration from traffic above. (See Sections 17.29 and 21.32.)

The amount of cribbing required will depend on the condition and type of soil. It is always better to place some cribbing at a questionable point than to risk a blow-in of earth or falling rock. (See Section 19.16.) A blow-in of earth means that as the excavating (heading) proceeds, loose or water-bearing earth flows or slides into the tunnel. The volume may be only a couple of cubic yards or it may be enough to fill several feet of the tunnel. At the first sign of movement of the soil, get the workers and equipment back a safe distance until the movement stops. If only a few cubic yards come in, it can be removed carefully, and liner plates can be installed. The void behind the plates can be filled with grout later. If a large volume of soil enters the tunnel, it may be necessary to stabilize the area. A large volume may also cause settlement in the surface above. There are several methods of overcoming a blow-in. The soil may be stabilized by injecting chemicals, or if the depth is not too great, sheet piling can be installed. An expensive way is to install air locks and proceed under compressed air until the area is bypassed. There was once a case where a blow-in occurred in a deep tunnel and a bulkhead had to be installed to seal the tunnel. The direction of the tunnel had to be changed to bypass the unstable area. Fortunately, the tunnel was for a large sewer, and the change in direction was not a problem. (See Figure 21.21.)

21.34. Turnbuckles

A turnbuckle is a device for keeping tension on a tie rod or cable. The device will have threads in each end which turn in opposite directions so that it will take up or release on both rods. Figure 24.16 indicates its use.

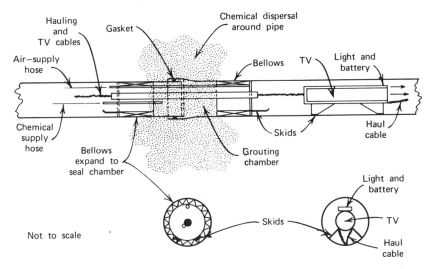

FIGURE 21.22. TV inspection of sewers. For inspection only, the TV cameras may be used without the chemical grouting unit. The sketch shows the sealing of a leak at a pipe joint. The same method is used to seal a leak at a crack resulting from a shear break. If there is any indication of leaking in the sewer line, it is customary to use the TV camera and the chemical grouting equipment together and seal the leaks as they show up during the inspection.

21.35. Turnkey Operations (Design Builds)

On turnkey operations, sometimes called design builds, the general contractor will confer with the owner on the type of structure required, and will arrange the financing, design the structure, and build it. When it is complete and ready for use, he will turn it over to the owner. If the structure is a building, the keys will be turned over to the owner, hence the name. Most states will require the contractor to have licensed engineers or architects do the design.

21.36. TV Inspection (Sewers)

TV inspection, as the name implies, is done by use of special TV equipment which is pulled through the sewer by a cable. A TV camera with a strong light will be connected to a TV set in a trailer on the surface. The cable will have foot marks to tell where the unit is located in the sewer pipe as it reveals leaks or other problems. A pressure grouting unit will be attached behind the TV camera, by which leaks can be injected with a chemical that is forced out into the earth outside the pipe and causes a gel to form that will seal the leak. This work is done by specially trained crews. (See Figure 21.22.)

22

22.1. Undercutting

Sometimes it is necessary to excavate back under an existing structure to install new work, or it may be necessary to excavate back under the side of an excavation to install the footing of a new structure. Such excavating is called undercutting. Figure 22.1 indicates the procedure.

22.2. Underground Obstacles

When construction is carried on within a built-up area, underground obstacles such as sewers, water mains, and telephone and power cables may be encountered. If the location of the obstacles is known, that information will be on the plans with as much data as are available. Sometimes the plans will note that the data are only approximate. The obstacles may be in use or abandoned, and the plans will usually state if they must be kept in service and if a bypass will be required to keep them in service. The method of payment may or may not be stated. If your crew uncovers an obstacle, known or unknown, you should inform your superior and keep an accurate record of any delay, stating why you were delayed and what was done to correct the problem.

None of the obstacles uncovered should be disturbed until the proper authority has had a chance to investigate and inform you if the item is in service and, if not, how it can be abandoned.

Sewers may be bypassed by pumping around the work, water mains can be piped around the work, cables can have temporary spliced bypasses (see Section 13.9).

374

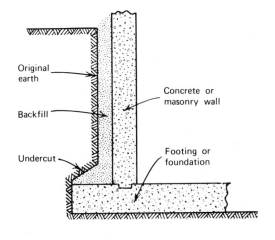

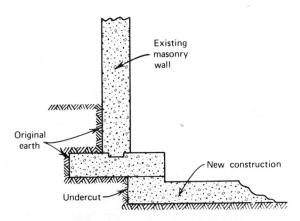

FIGURE 22.1. Examples of undercutting.

22.3. Underpinning

Figure 22.2 shows what underpinning is. Excavating under a reinforced concrete wall usually causes no problems, but masonry walls may develop cracks, depending on their condition. In either case, it is best to keep the width of the excavation as narrow as possible. One way to stabilize a masonry wall is to place a heavy timber or structural-steel channel on top of the footing and tight against the wall, with bolts through the wall at about 2-ft intervals.

The underpinning footings and columns should be poured with quick-setting concrete to about ½-in. below the bottom of the wall footing. This space is to be filled with expanding grout after the concrete has set.

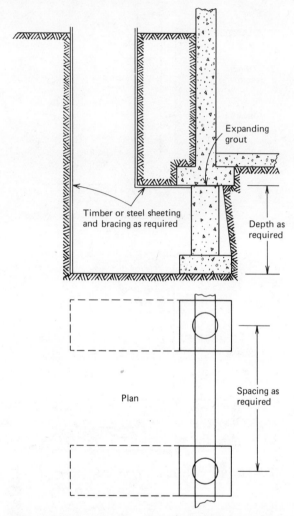

FIGURE 22.2. Excavation for underpinning an existing building.

If the wall is stabilized as noted above, it is best to leave the material in place. Forms for the column will usually be left in place.

22.4. Union Activities

Most large construction contractors are unionized, although many are not. You as a foreman may have come up through the union ranks, and if so, you can usually retain your card while working as a foreman. However, foremen are part of management, and regardless of your personal feelings toward unions, you will have to become management oriented if you hope to advance to a higher position. Most successful contractors work in reasonable harmony with the unions unless or until they feel the union's demands are unreasonable, and then friction can develop. Regardless of your feelings for or against unions, it is advisable for you to remain as neutral as possible. The problems will be worked out between the officials on both sides, regardless of what you say, and any statement by you one way or the other could cause resentment by the losing side.

Sometimes the union and management cannot agree, and a strike is called. If there is real bitterness between the parties, there is bound to be resentment by the loser, even after things are settled and work is resumed. Again, it is advisable for you to remain as neutral as possible.

22.5. Unions (Pipe)

A pipe union is a device whereby two lengths of pipe can be connected and then disconnected if needed, without removing the pipe. One piece of the device is threaded on one length of pipe, and the other half of the device has a threaded collar. The two pieces are screwed together, making a water-tight joint.

22.6. Unit Prices

There are two ways of paying for civil engineering construction, namely, by "unit price" or by "lump-sum" (see Section 13.26). Under unit price payment, the contractor is paid a stated amount for each item of work in terms of dollars per cubic yard, per linear foot, per square yard, or per pound, and so on, depending on what the unit is. Trenching will be by the linear foot for various depths; backfill will be by the cubic yard in place; special backfill will be by the cubic yard; embankment will be by the cubic yard in place or by the linear foot for a specified cross section; concrete is usually by the cubic yard, with the reinforcing steel being by the pound; and pavement may be by the square yard or by the linear foot for a specified width.

You must know how each unit of work is being paid for and keep accurate records of how many units you accomplished each day. You should retain the calculation used to determine what was done so that a check can be made with the engineer's quantities to see that the contractor gets paid for all that was done.

22.7. Unsatisfactory Work (See Section 19.6.)

22.8. Unstable Earth (See Section 20.61.)

23

23.1. Valves

Valves will be installed by plumbers, who will work under their own foremen; however, a construction foreman should have some general knowledge on the subject.

There are several types of valves, such as gate, butterfly, check, plug, pressure-release, and vacuum-release valves. The gate valve has a gate within the casting that moves open or closed on a threaded stem. The gate is machined to fit tightly within grooves in the housing and can stand very high pressures of liquid or gas. Butterfly valves have overlapping leaves that act quickly, but are not completely airtight. They are used on low-pressure lines for light liquids or air. The check valves are used to check a reverse flow in a pipe or conduit. When these valves are used on large pipe or conduit like culvert pipe, they are called flap valves. The plug valve is a heavy-bodied unit with a plug as a gate. The plug will have an opening through which heavy liquids, such as sewage sludge, can pass when the opening is parallel with the centerline of the pipe. The valve is shut off by turning the plug 90 deg so that the opening is at a right angle to the centerline. These valves are hard to operate and must be well lubricated. Pressure-release valves are placed at the low points in a force main on a sewer line or on a high-pressure water main. If the pressure in the pipe rises above what is considered safe, the valve will open and release the pressure. As soon as the pressure returns to a safe range, the valve will close. The vacuum-release valve works in just the opposite way. When a vacuum builds up in a line, the valve will open and relieve the pressure, and then close as the vacuum is relieved. Both such valves are set at the desired pressure by tightening or releasing a spring and bolt.

If you are assisting in the installation of valves, see that they are in operating order and that the inside is free of foreign matter such as sand, gravel, or small sticks, each of which could keep the valve from closing tightly.

23.2. Vapor Barriers

Most specifications will require that a vapor barrier be placed under concrete slabs poured on the ground and on the exterior surface of basement walls.

Those barriers under slabs will be plastic sheets laid on the ground before any reinforcing steel or concrete is placed. Care must be taken to see that the sheets are not punctured or torn as the workers place the steel and the concrete. One way to protect the sheets is to lay planks on them for workers to walk on until some concrete has been poured.

The barrier on exterior walls will usually be an application of hot asphalt sprayed or brushed on the dry wall and covered with a fabric, followed by another application of asphalt. Earth-filled arch bridges will usually have the same type of barrier placed over the slab before the earth fill is placed. The wall or bridge slab must be absolutely dry or the hot asphalt will not stick to it.

23.3. V-Belts

V-belts, like those used in the motors of automobiles, have sloping sides that fit tightly in the groove of a pulley and reduce the slippage.

23.4. Venturi Meters

Figure 23.1 shows one type of venturi meter; others are similar. Some have glass tubes and others have gauges to indicate the pressures, from which the flow of liquid is determined.

The meter must be set level and installed so that the flow will be in the right direction. Most meters will have an arrow on the side to indicate the direction of flow.

23.5. Vibro-Rollers (See Section 19.17.)

23.6. Vibrations

Civil engineering projects are concerned with two kinds of vibration: first, those caused by the operations on the job site, and second, those used to consolidate materials (see Figure 6.15).

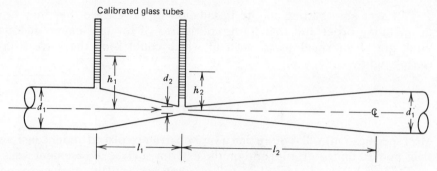

FIGURE 23.1. Venturi meter.

The first type of vibration is caused by blasting, by heavy equipment moving about the site, or by pile driving. Such vibrations can rattle windows, cause ground tremors that can be carried into nearby structures, and send shock waves through the air that, while causing very little damage, can be annoying.

Heavy earth-moving equipment operating near a residence can cause vibrations that seem to be severe, while it is really the sound that is severe, rather than the vibrations. You can calm a resident by placing a glass, filled to the very top with water, on a table, and it will show very little, if any, movement. If there are serious vibrations, waves will form on the water and might even spill some. If there are serious vibrations, attempts should be made to lessen or eliminate them.

Blasting operations in built-up areas are regulated by state laws, which should be carefully observed. If there is to be heavy blasting, it is recommended that the occupants of nearby structures be notified in advance and that the structures be examined for existing cracks. If existing cracks are found, they should be recorded and discussed with the owner. You might have problems with owners trying to collect damages for cracks that existed before your operations.

Pile driving can cause vibrations in an area, particularly if the soil is dense. However, it is the noise that causes most of the annoyance, and there is not much you can do about that. There are vibrating machines that sink the piles without driving, and these, of course, are more quiet.

The second kind of vibration is caused by the operation of equipment intended to consolidate materials such as concrete, earth embankment, subgrade, and bituminous paving. Such vibrations are not likely to disturb the neighbors. These vibrators are those inserted into a freshly poured concrete mass to consolidate it, those with steel plates which vibrate on the surface of a subgrade or embankment, and those that are part of the various types of steel-wheeled rollers.

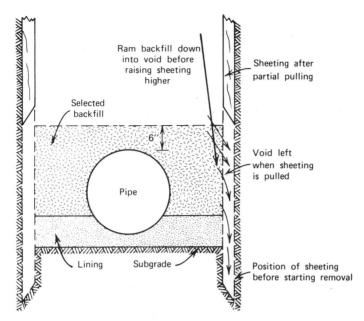

Ram backfill down
into void before
raising sheeting
higher

Sheeting after
partial pulling

Selected
backfill

6'

Void left
when sheeting
is pulled

Pipe

Lining   Subgrade

Position of sheeting
before starting removal

FIGURE 23.2. Filling void left when sheeting is pulled.

23.7. Visitors

Visitors, official or unofficial, should report to the superintendent and discuss any problems with him. Sometimes the contractor or one of his office men will come on the site and talk to the foremen just to be friendly or maybe to check on you or the superintendent. Unofficial visitors should be encouraged to talk to the superintendent so as not to hold up your work. When any official talks to you, be courteous, tell the truth, and do not try to overly impress anyone.

23.8. V-Notch Weirs

A V-notch weir, as the name implies, is in the form of a "V," as shown in Figure 24.18.

The weir may be at the end of a rectangular tank or around the perimeter of a circular one.

23.9. Voids

You and your crew should always be on the alert to eliminate voids as much as possible, as they can be costly to repair. Voids in concrete (honeycomb) can be avoided by proper vibrating (see Section 9.9). Voids in fill can cause settlement at the surface, requiring repair of lawns, walks, and driveways. Voids caused when sheeting is pulled can cause settlement of a pipe (see Figure 23.2).

Many voids do not show up until several months after the work is completed, and the contractor has to move back to the site to make repairs.

24

24.1. Wakefield Piling (Sheet)

Wakefield piling is used where tight, strong sheeting is needed. The piling is made by nailing three boards together as shown in Figure 24.1. The tongue and groove give strength and watertightness. Such sheetpiling will be hard to pull, as the ground moisture will cause the wood to swell. Such piling is seldom used today due to the high labor costs.

24.2. Walkie–Talkies

The walkie–talkie is a useful tool when concrete or material trucks have to be scheduled. It is also useful in calling for help in case of accident.

24.3. Wall Castings

Wall castings are used where pipes, conduits, or cables must pass through a wall or floor. (See Figure 24.2.) It is important that the centerline of the casting be exactly the same as the pipe or conduit; otherwise, a "Dutchman" or pipe bend will be required (see Section 5.27). The bell-and-spigot type will be caulked with jute and hot lead, the flange type will have gaskets, and the sleeve type will be caulked, as noted in the figure.

The castings must be carefully checked before the concrete is poured, and they must be secured so that they will not move during the pour.

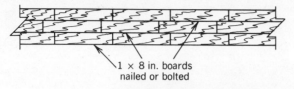

1 × 8 in. boards
nailed or bolted

FIGURE 24.1. Wakefield piling.

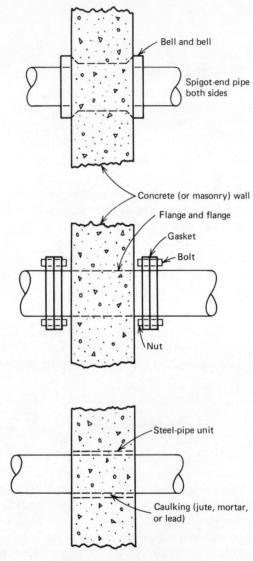

Bell and bell

Spigot-end pipe
both sides

Concrete (or masonry) wall

Flange and flange

Gasket

Bolt

Nut

Steel-pipe unit

Caulking (jute, mortar,
or lead)

FIGURE 24.2. Wall castings.

384

24.4. Water–Cement Ratios (See Section 4.35.)

24.5. Water Mains

Pipes for water mains can be of cast iron, reinforced concrete, asbestos–cement, or steel. The metal ones may be cement lined. The smaller sizes will be of cast iron or asbestos–cement. The smaller sizes will be laid at a more or less constant depth, following the contours of the earth. The larger sizes will usually be laid to a definite grade by use of a laying schedule (see Section 13.11).

The engineer or his inspector will require that the pipe be free of all foreign matter, pressure tested, and sterilized before acceptance, so it is

FIGURE 24.3. A 36-in. reinforced-concrete water main. Note that sections on a curve are cast to fit the horizontal and vertical angles. The pipe was laid by use of a laying schedule. Each section of the pipe was marked to correspond to the markings on the plans that gave alignment and grade. Note the "come-along" on top of the near pipe, which is to pull the sections tightly together. The white band around the pipe is a diaper used to grout the joint.

FIGURE 24.4. Control valves for a 24-in. water distribution system. The valves are horizontal with gear boxes at the side. The square nuts on the top of the gear boxes will have 6-in. cast-iron pipes, rising up to the ground surface so that a long-shafted key can be used to operate the valves. The valve on the right-hand side is temporarily plugged and will be used for a future extension. Note the dresser couplings on the pipe. The cross at the intersection and the valve housings will be encased in concrete.

FIGURE 24.5. These are 36- and 24-in. control valves in a valve pit just below the dam of a water storage reservoir. The vertical cylinders with the tie rods to the covers are part of a hydraulic mechanism for operating the valves. Note the manhole steps cast into the concrete wall. When the concrete roof is placed over the pit, a cast-iron manhole frame and cover will be set above the steps for access to the pit. The dark color around the pipe where it goes through the wall is recently placed nonshrinking grout to make the joint watertight.

wise to keep the pipe as clean as possible before and during laying. As each length is prepared for placement in the line, it should be inspected, and any substance inside should be removed before laying. Rodents will crawl into the pipe during wet or cold weather, so a light bulkhead should be placed over the end of the last pipe laid for the day.

All joints should be cleaned and lubricated just before laying, and the pipe should be shoved tight.

The plans will show the location and position of all tees and valves. These too should be clean when installed, and care should be taken to see that no small stones or gravel get in the valve where they could damage the gate and prevent it from fully closing. The pipe from the valve to the ground surface must be vertical over the nut so that the key will fit properly.

Cast-iron pipe can have very fine cracks that are not visible as the pipe is laid, but which will open under pressure and leak (see Section 4.9). See Figures 24.3–24.13.

FIGURE 24.6. One possible pipe arrangement in a pumping station just below the dam of a man-made lake. The canvas covers an electric motor for the centrifugal pump. The unit with a horseshoe-shaped top is a check valve to prevent the water from flowing backward when the pump is stopped, as the reservoirs along the main are higher than the surface of the lake. Note the "increaser" unit between the pump and the check valve, the pipe connecting to the main being 2 in. larger than the supply line from the lake.

At many small pumping stations such as this, the contractor will complete all of the building except the roof and then install the pipe and equipment by use of a crane from outside the building. When the installation of all pipes and equipment has been completed, the roof will be placed.

At this location, there were two identical pumps. The check valve of the other pump can be seen in the foreground.

FIGURE 24.7. A distribution point in a 30-in. water main. The main is reinforced-concrete pipe, except at pumping stations and distribution points. The two valves on the left are for future connections to lateral mains. Note the two ¾-in. steel rods in the foreground that hold the joint between the steel and the concrete pipe. Without such ties, the water pressure would force the joint apart. Note that the valves lie horizontally. The 2-in. square nut on the end of the valves will have a 6-in. pipe set vertically above it that will extend to the ground surface. Through this pipe, a large T-wrench will be lowered to the nut for operating the valve.

FIGURE 24.8. One method of repairing a 30-in. concrete-pipe water main. In this case, a 6-in. hole was sealed by use of the device shown. The surface of the pipe was scrubbed clean with a wire brush; then a rubber gasket and the steel plate, held in place by the four U-bolts, were installed. Sometimes the plate and the bolts are covered with a bituminous paint to resist rusting.

388

FIGURE 24.9. The installation of a 30-in. steel-pipe water main. The pipe is wrapped with a special moisture-resistant bitumen-impregnated fabric. The pipe is quite flexible, so that several 40-ft lengths can be set on timbers above the trench, welded together, and then lowered into the trench by crane.

FIGURE 24.10. The wrappings are removed for a couple of feet at the ends of each section so the welding will not burn the wrapping. After the weld has been completed and inspected, the pipe will be rewrapped. The joint will be welded both inside and out. The inside weld will be rather difficult, requiring a small welder and forced air ventilation.

FIGURE 24.11. After a weld joint has been completed, it will be cleaned by use of a powered wire brush.

Before the joint is rewrapped, it will be x rayed and the film will be examined for flaws.

FIGURE 24.13. Another view of a pipe being installed. The foreman looking into the pipe is conferring with a worker at the last joint where he will check to see that the joint is up tight and that the gasket did not get pushed out of place as the joint was closing. He will also place mortar in the joint to provide a continuous smooth surface inside the pipe.

FIGURE 24.12. After the joints have been rewrapped and all spots where the wrapping has been damaged have been repaired, the pipe is "rung." The "ringing" is done with the device shown. A steel wire collar is placed around the pipe and rolled along by pushing with the handle shown. A wire through the handle connects the collar with a sounding device shown on the ground. If the wrapping is not equal to or more than the minimum thickness required, a bell will ring and the spot will be marked. After repairs, the spot will be rung again.

Note in the background that the crane is lifting the pipe by use of a canvas sling. The workers remove the timber supporting the pipe, and it is lowered into the trench. Wrapped pipe such as this must be handled very carefully. It is usually bedded in sand or other fine material to prevent the wrapping being cut or punctured.

24.6. Water-Plug

Water-plug, like Sika, is a trade name for a chemical used to stop water leaks in structures. (See Section 13.14.)

24.7. Water Stops

An expansion or construction joint in a concrete wall of a tank containing water or other liquid, or in a concrete wall of an underground structure in an area of high groundwater, will leak without a water stop. These stops were originally made of copper or aluminum, but most are now of rubber or a composition. Figure 24.14 shows how they are placed. The stops in the wall must connect with the stop along the base of the wall at the foundation. The plans will indicate the method of connection. Unless special care is taken, the stops can be damaged when removing the forms between pours.

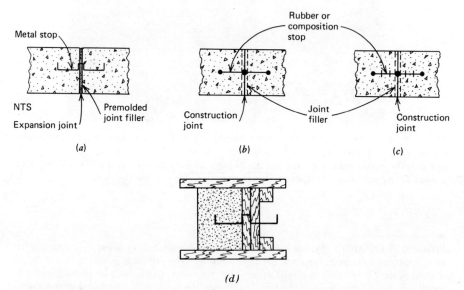

FIGURE 24.14. (a) An expansion joint with a copper or aluminum water stop with premolded joint filler. The size and gauge of the water stop and the thickness and material of the joint filler will be given in the specifications. When the concrete is being poured on one side of the joint, the exposed half of the stop must be protected from being crushed or bent out of shape, or the joint may leak. (b) One design of a rubber or composition stop. The sketch shows the stop in a construction joint, but it can be used in an expansion joint by adding premolded filler, as noted by the dotted lines. (c) Another design of the stop. The fins increase the resistance to leaks. (d) A method of setting forms around a water stop. When interior forms are removed, the stop will be ready for the next concrete pour.

One way to keep the stop in place while making the second pour is to punch a small hole every few feet in the outer edge of the stop and tie it to the reinforcing steel with light-gauge wire.

24.8. Water Towers

A water distribution system will have one or more water towers on the line to help keep the pressure up and to provide storage for extra water to fight fires.

The tower, including the tank and the legs, will be erected by steelworkers with their own foreman, but the concrete foundations for the legs will be poured under a construction foreman.

The layout of the foundation, and particularly of the anchor bolts, must be exact so that the legs can be positioned as required. When the anchor bolts are in position, but before the concrete is poured, the position should be rechecked, and the bolts should be secured so that they cannot be moved during pouring. The length of the bolt above the concrete must be enough that the baseplate and nut can be installed. See Figures 24.15 and 24.16.

24.9. Weather

Weather can affect the progress of the work, and thus the costs. Sometimes the weather conditions can be the basis of a claim for extra payment, so you should keep an accurate record of any adverse weather and delays it caused your crew.

FIGURE 24.15. A water tower and a closeup of a footing. High tanks provide pressure and extra water in case of fire. The legs are set on concrete footings and held in place by two 1-in. anchor bolts. Note diagonal tie rods. The tension is applied by a turnbuckle in each rod.

FIGURE 24.16. Looking up at the underside of an elevated water storage tank under construction. The dark lines are the welds, which have been recently painted. This 750,000-gal tank will supply line pressure and water for fires at the end of an extensive distribution system. Note that the crossed stabilizing rods have turnbuckles to regulate the tension.

24.10. Weather Stripping

Weather stripping will usually be installed by the carpenter when she installs the doors and windows; however, the job may be assigned to you. The stripping should be installed for the full length of the sides, top and bottom, except that the door will have a threshold at the bottom. There should be enough tension on the stripping that it will fit tightly against the door or window.

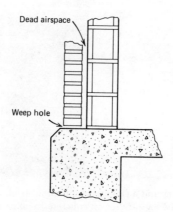

FIGURE 24.17. Weep hole.

24.11. Weep Holes

Weep holes in structures are used to allow the moisture or groundwater that accumulates to drain off. Figures 19.8 and 24.17 indicate how they are used.

24.12. Weirs

On civil engineering projects, weirs are used for different purposes. A weir for recording the flow of water will be with or without end contraction. There are charts available for calculating the flow at various depths.

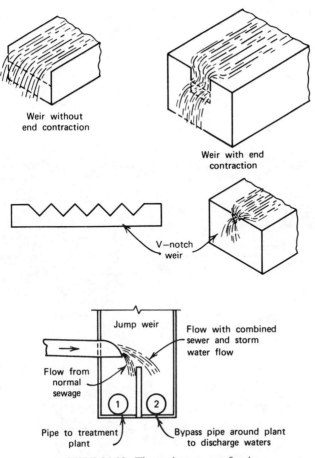

Weir without
end contraction

Weir with end
contraction

V—notch
weir

Jump weir

Flow with combined
sewer and storm
water flow

Flow from
normal
sewage

1   2

Pipe to treatment
plant

Bypass pipe around plant
to discharge waters

FIGURE 24.18. The various types of weirs.

Another type of weir is the V-notch. As the name indicates, the weir is a series of V-notches that control the distribution of the water during periods of low flow.

There is also a jump weir used in sewers to divert storm water. A study of Figure 24.18 will help explain the various types.

24.13. *Welding*

Welding has replaced riveting on almost all structural-steel erection. Welding is also used to connect steel plates for tanks to repair broken steel equipment and parts.

Require the welders to produce a certificate showing that they are qualified to do the class of welding required on the project. You yourself must be qualified to pass on the work involved. The welders become qualified by attending a welding school or by being tested in a local welding shop that is approved to certify. You may become qualified by doing the same thing. In this way, you can obtain enough knowledge to recognize a good weld.

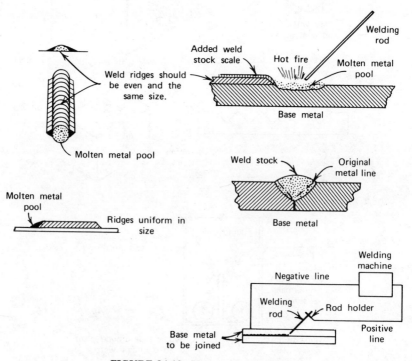

FIGURE 24.19. How welds are made.

Basically, you must see that the units to be welded are held firmly in place, that the weld ridges are uniform in width, height, and spacing, that the slag is chipped off, and that there are no breaks in the spacing of the ridges. (See Figure 24.19.)

24.14. Well Points

Well points are used to dewater an area of high groundwater. The system involves a series of 1½- or 2-in. pipes set vertically in the ground and attached to a 6–10 in. header pipe on the surface. The vertical pipes will have a 2-ft-long screen and a valve on the bottom. The valve allows water to pass out while the pipe is being jetted down, but will close so that the

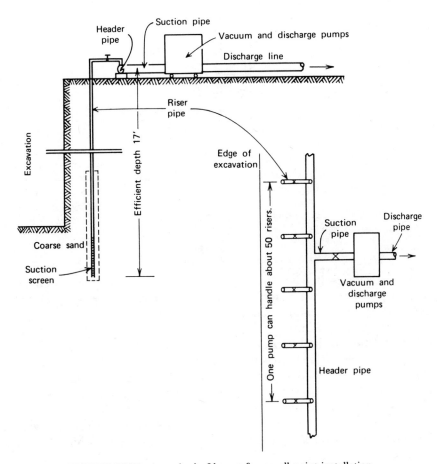

FIGURE 24.20. A method of layout for a well-point installation.

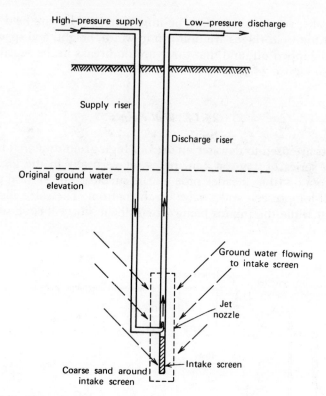

FIGURE 24.21. Well-point deep-water jet system. A typical installation. The jet system, like the standard well-point system, is manufactured by various companies. Each system varies, but the general principle is similar to that shown in the sketch. The intake screen is at the bottom of the discharge pipe, which is jetted into the earth. Coarse sand is used, unless the area is of very pervious material such as sand and gravel. The high-pressure jet causes a vacuum, which pulls the groundwater through the screen and then forces it up to the surface through the discharge pipe.

incoming water will pass through the screen. The water is raised by a vacuum produced by a pump attached to the header pipe. Figures 24.20 and 24.21 show the arrangement of the system.

When the vertical pipes are jetted down, a cast-iron "star" or a chain will be attached to the lower end to make the hole larger. After the jetting is complete, the star or chain will be pulled to the surface by a rope attached. The wider hole is to allow coarse sand to be deposited for about half the depth of the hole. This sand gives a greater drainage area and faster movement of the groundwater to the screen.

When jetting the vertical pipes, care must be taken in areas where there are power cables overhead. No pipe should come within 10 ft of a power cable.

Many contractors hire well-point crews to install and operate the system, while others own their equipment and install and operate it with their own crews.

The riser pipe will be 15–20 ft long and attached to a high-pressure pump by a length of hose. The riser is placed vertically on the ground after checking for overhead power cables. It usually takes two workers to handle the pipe and one for the hose. The riser will be placed close enough to the header pipe that it can be connected to the header pipe. As the jetting water is turned on, the riser will sink fast for the first few feet and, in most cases, will sink to the desired depth in a couple of minutes. The riser must be set so the top will be in position to connect to the header pipe. As soon as the riser is in position, the sand should immediately be poured into the hole around the pipe before any of the sides of the hole can collapse. Do not forget that the star or chain must be pulled out as soon as the riser is connected to the header. The jetting will usually result in a puddle of water on the surface, so the workers should have boots. Also, there will be a great deal of silt and fine sand washed up during the jetting, the volume depending on the porosity of the soil. The more clay there is in the soil, the more water will be on the surface. This water and silt can cause problems, and a way must be found to deal with it, depending on the location and the damage it could do.

After the risers are installed and connected, the vacuum pumps are turned on, the valves at the connecting pipes are opened, and the removal of the groundwater begins. At first, there will be fine silt in the discharged water, but this should clear up in a few minutes. In soil with low clay content, there can be a considerable amount of water discharged, so plans must be made as to where it will be disposed of.

24.15. Wells

When groundwater is encountered while excavating in pervious soil, a series of wells can be used to pump down the area if the soil is open enough to allow fast flow to the pumps; otherwise, well points would do a better job. (See Sections 5.6. and 24.14.)

In processing plants, permanent wells are sometimes required to supply cooling water, and if so, they will be detailed on the plans.

Most such wells will be constructed by lowering 2-, 3- or 4-ft-diameter concrete pipe in the same manner as for a caisson. Two or three feet of coarse gravel will be placed at the bottom to filter the water.

Sometimes an underground spring will be encountered in which there is considerable flow in a small area. In such a case, a well placed to intercept the flow and pump it off is an economical solution.

24.16. Whackers

"Whacker" is the trade name of one of several vibrators used to compact subgrade or fill. Whackers come in several sizes and are similar in design. They will have a heavy steel plate, on which is mounted an electric or gasoline motor which vibrates the plate. There will be a handle similar to that on a lawn mower for control on the unit.

24.17. Wharves

A wharf, also called a dock, runs parallel with the shoreline, as compared to a pier, which juts out into the water. Both are facilities for the loading and unloading of ships. They can range from a simple timber platform to a concrete and steel structure with an industrial railway, an overhead gantry crane, and storage warehouses. Such structures will involve timber, concrete, structural steel, and piling, all of which are explained elsewhere in this book. (See Section 17.14.)

24.18. Wide Flanges (Steel)

The wide flange is similar to the structural-steel I-beam, except the flanges are wider, giving the unit greater strength in both directions and thus making it useful as a column as well as a beam. (See Figure 20.32.)

24.19. Winches

A winch is a revolving drum on which rope or cable is wound. It is used to raise the anchor on large ships and on construction hoists. It may be hand or power operated.

24.20. Windows

Windows, doors, and louvers are installed by the carpenters, although the metal types may be handled by ironworkers.

Wooden units should have their backs well primed, especially the ends with open grain, before they are installed. Metal units should come with a prime coat of paint on all sides. All units should fit closely into the opening to reduce the amount of caulking. The double-hung units should move up or down freely and stay in position. Casement windows should swing easily on their hinges, and all units should fit tightly when closed, in order to exclude the weather and not rattle.

The windows will come completely assembled, ready to be placed in the opening. Care must be taken to see that, while they fit closely, they do not bind. Strips of paper or whitewash should be placed on the glass to indicate its presence and avoid damage.

24.21. Winter Construction (See Section 4.43.)

24.22. Wire Mesh

Steel wire mesh is used as reinforcement in light concrete slabs, with stucco or Gunite, and as reinforcement for concrete highway slabs. The mesh is also used as guard fencing around machinery or openings in floors.

24.23. Witnessing

You may be called when there is a disagreement between the contractor and the owner that cannot be settled without legal action or arbitration proceedings. If called under either circumstance, take to court the records called for without any attempt to withhold some. Tell the truth when asked a question, but do not volunteer any information not specifically asked for. When asked a question by the opposing attorney, delay your answer for a few seconds to give your attorney a chance to object if he so wishes. It is also a good idea when on the witness stand to make it a practice to delay answering for a few seconds to give yourself a chance to consider your answer to a tough question.

It is a good practice to speak slowly and distinctly from the start. If you give a quick, snappy answer to one question and delay on others, the opposition and the judge may think you are coloring some of your answers.

Sometimes the opposing attorney will try to confuse you or get a favorable reply by instructing you to answer yes or no to his question. If you can give a satisfactory yes or no answer, do so. However, if you feel that such an answer could be misleading, you should tell the judge, who will probably tell you to answer in your own words. If your attorney is alert, he will probably object to the simple yes or no, but in any case, you must follow the judge's instructions.

When disagreements between contractor and owner must be settled in court, you will probably be called to attend an examination before trial (EBT). At such a hearing, each attorney tries to find out as much as he can about what evidence the opposing attorney has. If he decides the other side has arguments that outweigh his, he usually will try for an out-of-court settlement.

At an EBT, you will be under oath just as if you were at a trial, and you should conduct yourself in the same manner. Remember, tell the truth, the whole truth, and nothing but the truth, no matter who it hurts.

Accidents on the job site may be investigated by OSHA or other authority. If you are questioned by any of these, conduct yourself as suggested above and immediately inform your superior if he does not already know.

Different witnesses to an accident may give different testimony which conflicts with what you claim happened. In such a case, it is recommended that you not be swayed by their views unless they convince you that you are wrong. It is best to make a few notes immediately after an accident, outlining what you believe you saw, and you should refer to those notes in private before testifying.

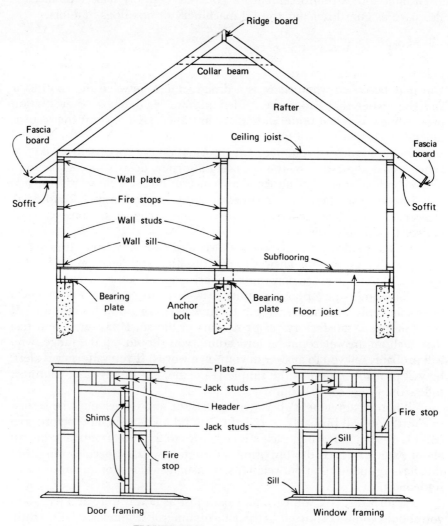

FIGURE 24.22. Frame construction.

24.24. *Women Workers (See Section 7.3.)*

24.25. *Wood Construction (Frame)*

Civil engineering projects seldom use "frame" construction, although some industrial and commercial structures will have frame partitions.

The following will give you some basic knowledge on this type of construction. Heavy solid timber construction is rare today, although laminated sections made by gluing wood planks together are common. The laminated units may be straight like a beam or arched. They are common in gymnasiums and other sports arenas. These heavy units will usually have cast-iron bearing plates. The plans will show the type and how it is anchored. Frame partitions will be erected as shown in Figures 24.22 and 24.23 and will be nailed together.

24.26. *Wooden Pipe*

When water is transported from a reservoir to a power plant to turn the generators, large-diameter pipe is used. It may be of steel, reinforced concrete, or wooden slats.

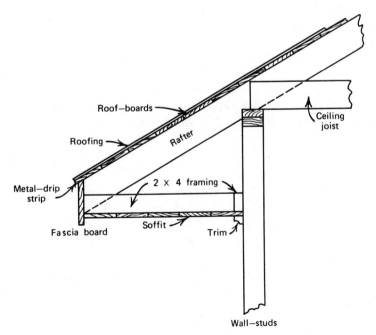

FIGURE 24.23. One method of constructing a cornice or overhang of a roof. The details will vary from job to job.

When the pipe is of wood, tongue-and-groove slats will be placed lengthwise and held in place by heavy steel rods. This type of pipe was popular years ago, but is seldom used today because of the high cost of labor and materials.

24.27. Work Areas

In an open field, there is little worry about the work area, but built-up, confined areas can cause problems. It is not uncommon for property owners along the line of work to make claims for damage when there really was none or to make exaggerated claims for minor damage in the hope of getting some extra money. You must know the limits of your work area and keep your operations within those limits if possible. Should damage occur, make an honest report of what happened, how the damaged occurred, and exactly what was damaged and to what extent. Some of the things for which owners make claims for damage are: vehicles driving over the lawn, grass killed due to materials being stored on it, trees or bushes damaged by moving equipment, sidewalks or driveways damaged by vehicles, trees killed by root damage due to trenching, and plaster cracked due to vibration from blasting or movement of heavy equipment.

If some damage does occur due to your operations, it is best to let your superior deal with the complainant.

24.28. Worker's Compensation

Worker's compensation is insurance for the worker should he be injured or become ill due to working conditions on the job site. The requirements vary from state to state and will be spelled out in the contract documents.

FIGURE 24.24. Wye for bell-and-spigot pipe.

As with any insurance, there will be forms to be filled out detailing what happened and why, with the names of witnesses, if any. These papers should be carefully and truthfully filled out, as they can affect the award of compensation some time later.

24.29. Work Schedules (See Section 4.60.)

24.30. Wrought Iron

Wrought iron is used to make pipes, structural columns, and grillwork such as ornamental fencing. The metal is tough but malleable and resists rust.

24.31. Wyes

A wye is a Y-shaped pipe section used when two pipes are to be connected, but not at a right angle. (See Figure 24.24.) (See Section 21.6.)

25

25.1. X-Braces

X-braces are used in structural framework to provide stiffeners for both directions. They are used between bridge trusses to keep the trusses vertical and to resist wind pressure. They are used in the walls of high buildings to resist wind pressure. The braces may be of structural steel or reinforced concrete.

25.2. X rays

The engineer or his inspector may require x rays to check welding or examine for flaws in steel. An x-ray specialist will be employed to do the work and interpret the findings.

25.3. X-Section

The term X-section is an abbreviation of the words "cross section." (See Section 4.62.)

26

26.1. Yard (Area or Volume)

When some work on the job is to be paid for "by the yard," it is necessary to know if it is "by the square yard," which is area, or "by the cubic yard," which is volume. The specifications will indicate which is intended.

26.2. Yard (Construction)

The layout of a construction job yard can affect the efficient operation of the job. The following layout is recommended:

1. The site should be well drained, or the movement of machines will quickly turn it into a sea of mud. If there is no natural drainage, it should be provided by runoff ditches to get the rainwater off the site as quickly as possible.
2. The job office should be located where a good view can be had of as much of the work as possible. If there will be work after dark, floodlights should be placed so as to illuminate all areas where work will be carried on.
3. Storerooms or warehouses should be placed to provide easy access for the trucks; if possible, the platform should be at the level of the truck body.
4. The toolshed should be a strong building with lock and key.
5. There should be a designated area for lunch which can be quickly cleaned to keep flies and rodents away.

6. Construction materials should be stored as close to operations as possible without interfering with the movement of equipment.

7. Finally, the job superintendent will indicate how far you should go in setting up the site, the cost of which will be included in the project estimate.

26.3. Yarn

Yarn is used to make a braid for the joints of bell-spigot pipe. (See Section 11.11.) The yarn may be treated or untreated as required.

26.4. Yees (See Section 24.31.)

26.5. Y-Levels

Some surveyor's and builder's levels are what is known as the Y type, which means the telescope is held in place by two Y-shaped supports. The "Y" can be opened and the telescope can be removed. The other type of level is the "dumpy" level, in which the telescope is cast solid with its supports and cannot be removed. Both are similar in operation, except that care must be taken to see that the telescope in the Y type is in the correct position.

27

27.1. Z-Bars

Z-bars are structural shapes used to support brick or other exterior facing on a building. (See Fig. 20.32.)

27.2. Zoning

Zoning laws can affect the operation of a project. They specify the types of structures that can and cannot be erected in an area and may restrict the time and extent of operations.

Appendix A

The Metric System Conversion

Inches to millimeters	1 in. = 25.4 mm
Millimeters to inches	1 mm = 0.0393700 in.
Feet to meters	1 ft = 0.3048 m
Meters to feet	1 m = 3.28084 ft
Yards to meters	1 yd = 0.9144 m
Meters to yards	1 m = 1.0936 yd
Miles to kilometers	1 mi = 1.6903 km
Kilometers to miles	1 km = 0.62137 mi
Pounds to kilograms	1 lb = 0.45359 kg
Kilograms to pounds	1 kg = 2.2046 lb
Quarts to liters	1 qt = 0.94632 L
Liters to quarts	1 L = 1.0567 qt
Gallons to liters	1 gal = 3.7853 L
Liters to gallons	1 L = 0.26418 gal
Square inches to square centimeters	$1 \text{ in.}^2 = 6.4516 \text{ cm}^2$
Square centimeters to square inches	$1 \text{ cm}^2 = 0.15500 \text{ in.}^2$
Square feet to square meters	$1 \text{ ft}^2 = 0.09290 \text{ m}^2$
Square meters to square feet	$1 \text{ m}^2 = 10.7639 \text{ ft}^2$
Square yards to square meters	$1 \text{ yd}^2 = 0.83613 \text{ m}^2$
Acres to hectares	1 ac = 0.40469 ha
Hectares to acres	1 ha = 2.4711 ac
Cubic inches to cubic centimeters	$1 \text{ in.}^3 = 16.38706 \text{ cm}^3$
Cubic centimeters to cubic inches	$1 \text{ cm}^3 = 0.06102 \text{ in.}^3$
Cubic feet to cubic meters	$1 \text{ ft}^3 = 0.02832 \text{ m}^3$
Cubic meters to cubic feet	$1 \text{ m}^3 = 35.3147 \text{ ft}^3$
Cubic yards to cubic meters	$1 \text{ yd}^3 = 0.76464 \text{ m}^3$
Cubic meters to cubic yards	$1 \text{ m}^3 = 1.30795 \text{ yd}^3$

Complete conversion tables can be obtained in most local bookstores.

410

Appendix B

Average Weights of Materials

Concrete: reinforced—150#/cf—4050#/cy
              unreinforced—140#/cf—3780#/cy
Run-of-bank gravel: 2650#/cy
Crushed stone: 2700#/cy
Sand: 3000#/cy
Brick: common—5½# each
      face—6½# each
Blocks, concrete: 38# each
Bituminous concrete: 2 ton/cy
Timber: 40#/cf
Pipe: Asbesto cement (A–C) weight in pounds per linear foot (plf)

Size	2400 Class	3300 Class	5000 Class
8 in.	13	14	none
10 in.	18	21	26
12 in.	23	27	34
16 in.	35	41	51
18 in.	42	49	61

Pipe, concrete: 8 in.— 49#/lf
               10 in.— 78#/lf
               12 in.—100#/lf
               16 in.—145#/lf
               18 in.—185#/lf

Appendix C

Useful Calculations

There are a few calculations that a foreman should be able to make. All require only a knowledge of arithmetic.

Referring to Figure C.1A, the area of a rectangle surface is: 25 ft × 6 ft = 150 ft^2, 150/9 = 16.67 yd^2.

Referring to Figure C.1B, the area of a trapezoidal surface is (20 + 18)/2 ft × 5 ft = 95 ft^2 and (22 + 20)/2 ft × 5 ft = 105 ft^2. Assuming these two areas are the vertical sections in an excavation and are 50 ft apart, what is the volume of the excavation? Get the average of the two sections, which will be (95 + 105)/2 = 200/ 2 = 100 ft^2. The 100 ft^2 × 50 ft = 5000 ft^3. The 5000 ft^3/ 27 = 185.19 yd^3 = volume of the excavation.

Referring to Figure C.1C to calculate the area of an irregular surface as shown, divide the area into rectangular and triangular sections. Then, section

$$(1) = 10\,\text{ft} \times 10\,\text{ft} = 100\ \text{ft}^2$$

$$(2) = 12\,\text{ft} \times 5\,\text{ft} = 60\ \text{ft}^2$$

$$(3) = (12\,\text{ft} \times 5\,\text{ft})/2 = \underline{30}\ \text{ft}^2$$

$$\text{Total } \overline{190}$$

Referring to Figure D, to calculate the area a cubic yard of material will cover at a specified depth, assume a cubic yard of concrete to be spread to a depth of 6 in. A square yard is 9 ft^2. A depth of 3 ft is 36 in. The 36 in. will make six layers 6 in. thick. Six layers at 9 ft^2 each equals 54 ft^2, and 54/9 = 6 yd^2. Thus, 1 yd^3 of concrete at 6 in. thick will cover 6 yd^2.

Similar calculations can be made by the same methods.

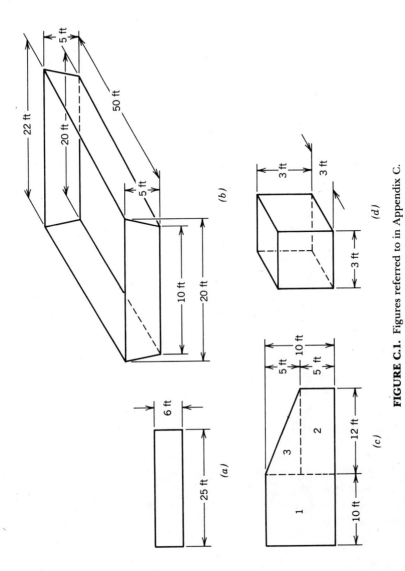

FIGURE C.1. Figures referred to in Appendix C.

Appendix D

Weights and Measures

Linear feet:

1 in.	=	0.0833	ft
12 in.		1.0	ft
3 ft	=	1.0	yd
16½ ft	=	1.0	rod
5280 ft	=	1.0	statute mile

Square measure:

144 in.^{2}	=	1.0	ft^2
9 ft^2	=	1.0	yd^2
43560 ft^2	=	1.0	ac
4840 yd^2	=	1.0	ac

Cubic measure:

1728 in.^{3}	=	1.0	ft^3
27 ft^3	=	1.0	yd^3

Metric conversion (approximate):

inches	×	2.5	= centimeters
feet	×	30.0	= centimeters
yards	×	0.9	= meters
miles	×	1.6	= kilometers

Note: Carpenters' rules and tapes are in feet and inches. Surveyors' rules and tapes are in feet and tenths of feet.

Appendix E

Temperature Conversion

A Fahrenheit reading = a Celsius reading multiplied by 1.8 plus 32:

$$f = 1.8c + 32$$

A Celsius reading = a Fahrenheit reading minus 32 divided by 1.8:

$$c = \frac{f - 32}{1.8}$$

Water freezes at 0°C and 32°F.
Water boils at 100°C and 212°F.

Author's Biography

JAMES E. CLYDE, a retired professional engineer and land surveyor, was an organizing partner of the Consulting Engineering firm of Barton, Brown, Clyde & Loguidice, P.C., where he was Vice-president in charge of field services and Chairman of the Board. He has supervised construction and/or designed numerous large projects, including bridges over the Mississippi, Illinois, Susquehanna, and several other rivers; overseas military bases; railroad yards; water supply and distribution; sewers and sewage treatment; highways; airports; and heavy earth moving. During World War II, he commanded an army engineering construction battalion in the South Pacific.

Mr. Clyde is an Honorary Member and past President of the Consulting Engineers Council of New York State, past Director of the Consulting Engineers Council of the United States, Fellow and Life Member of the American Society of Civil Engineers, and Life Member of the National and State Societies of Professional Engineers.